49

Advances in
Polymer Science

Fortschritte der Hochpolymeren-Forschung

Living Polymers and Mechanisms of Anionic Polymerization

By M. Szwarc

With 69 Figures

Springer-Verlag
Berlin Heidelberg New York 1983

Editors

Prof. Hans-Joachim Cantow, Institut für Makromolekulare Chemie der Universität,
Stefan-Meier-Str. 31, 7800 Freiburg i. Br., BRD

Prof. Gino Dall'Asta, SNIA VISCOSA – Centro Studi Chimico, Colleferro (Roma),
Italia

Prof. Karel Dušek, Institute of Macromolecular Chemistry, Czechoslovak Academy of
Sciences, 162 06 Prague 616, ČSSR

Prof. John D. Ferry, Department of Chemistry, The University of Wisconsin, Madison,
Wisconsin 53706, U.S.A.

Prof. Hiroshi Fujita, Department of Macromolecular Science, Osaka University,
Toyonaka, Osaka, Japan

Prof. Manfred Gordon, Department of Chemistry, University of Essex,
Wivenhoe Park, Colchester C04 3 SQ, England

Prof. Joseph P. Kennedy, Institute of Polymer Science, The University of Akron,
Akron, Ohio 44325, U.S.A.

Prof. Werner Kern, Institut für Organische Chemie der Universität, 6500 Mainz, BRD

Prof. Seizo Okamura, No. 24, Minami-Goshomachi, Okazaki, Sakyo-Ku, Kyoto 606,
Japan

Prof. Charles G. Overberger, Department of Chemistry, The University of Michigan,
Ann Arbor, Michigan 48 104, U.S.A.

Prof. Takeo Saegusa, Department of Synthetic Chemistry, Faculty of Engineering,
Kyoto University, Kyoto, Japan

Prof. Günter Victor Schulz, Institut für Physikalische Chemie der Universität,
6500 Mainz, BRD

Dr. William P. Slichter, Chemical Physics Research Department, Bell Telephone
Laboratories, Murray Hill, New Jersey 07971, U.S.A.

Prof. John K. Stille, Department of Chemistry, Colorado State University, Fort Collins,
Colorado 805 23, U.S.A.

ISBN-3-540-12047-5 Springer-Verlag Berlin Heidelberg New York
ISBN-0-387-12047-5 Springer-Verlag New York Heidelberg Berlin

Library of Congress Catalog Card Number 61-642

Table of Contents

Living Polymers
and Mechanisms of Anionic Polymerization

Michael Szwarc

Dept. of Chemistry, New York State College of Environmental Science, Syracuse, N.Y. 13210, USA

Following a short introduction and a brief discussion of thermodynamics of propagation, the kinetics and mechanisms of anionic polymerization are reviewed. The systems involving living polymers, a term introduced by this writer, are discussed in greater detail because the existence of various ionic species participating in polymerization was clearly revealed by their studies. Indeed, a large part of this review is concerned with the methods used in identification of these species and determination of their role in various polymerizing systems. The results of such studies are compared with those derived from investigations of radical-anions, carbanions and other ionic species.

Initiation and propagation of anionic polymerization are treated separately. Both homogeneous and heterogeneous initiation processes are discussed, including systems involving electron-transfer processes, zwitter ions, charge-transfer complexes, etc. Special attention is paid to lithium alkyl initiators in view of their complexity. The effects exerted upon the rate and stereo-chemistry of propagation by the nature of counter-ions, aggregation, solvents and solvating agents, temperature, pressure, etc., are thoroughly discussed.

I. Introduction

I.1. Early Developments in Anionic Polymerization

First reports describing the processes classified today as "anionic polymerizations" appeared at the end of the last century. At that time several authors reported the formation of gums and resins produced under the influence of alkali metals. Deliberate activities in this field began in the first decade of this century. A patent, issued in 1910 to Matthews and Strange[1], claimed the polymerization of dienes induced by alkali metals. In the following year Harries[2] published his pioneering study of isoprene polymerization, and three years later Schlenk[3] reported the formation of presumably high-molecular weight polymers prepared in ether solution by reacting styrene or 1-phenylbutadiene with sodium dust. In fact, the concept of macromolecules was not yet established in those days, hence the produced materials were described in terms borrowed from colloid chemistry.

The modern concept of linear polymers was introduced in 1920 by Staudinger, who fully recognized the idea of chain addition reaction yielding long molecules composed of monomeric units linked by covalent bonds. He was also the first to understand the anionic character of formaldehyde polymerization initiated by bases such as sodium methoxide[4]. In fact, studies of this reaction led him to the notion of linear macromolecules. Polymerization of ethylene oxide initiated by alkali metals and reported as early as 1878[5] could also be interpreted in these terms.

Further developments in the field of chain polymerization were centered on radical poly-addition. Its mechanism was firmly established in the 1930's and attracted much attention. The interest in anionic polymerization was marginal and the activities in this field were centered at that time around Ziegler in Germany and Lebedev in Russia. Both groups were interested in polymerization of styrene and dienes initiated by sodium metal and their work led to industrial production of synthetic rubber marketed by I. G. Farbenindustrie as "Buna".

Systematic studies allowed Ziegler to formulate the initiation step as the addition of two sodium atoms to a monomer with formation of two covalent Na-carbon bonds[6], e.g.,

$$CH_2:CH \cdot CH:CH_2 + 2Na \rightarrow Na-CH_2 \cdot CH:CH \cdot CH_2-Na .$$

The concept of carbanions and ion-pairs was at its infancy at that time. The isolation of such adducts eluded him. Nevertheless, he argued for their existence by demonstrating the formation of butene-2 in reaction performed in the presence of an excess of methyl-aniline[6a], a compound that does not react directly with sodium metal. Since two moles of sodium amide were obtained for each mole of the butene formed by this process, Ziegler described its course by the scheme

$$Na \cdot CH_2 \cdot CH : CH \cdot CH_2 \cdot Na + 2 PhNH \cdot CH_3 \rightarrow CH_3 \cdot CH : CH \cdot CH_3 + 2NaN(CH_3)Ph \ .$$

Our modern description would probably follow the line

$$CH_2 : CH \cdot CH : CH_2 + Na \rightarrow (CH_2 : CH \cdot CH : CH_2)^- \cdot, Na^+ \xrightarrow{PhNHCH_3} CH_3 \cdot CH : CH \cdot CH_2 \cdot$$

$$CH_3 \cdot CH : CH \cdot CH_2 \cdot + Na \rightarrow CH_3 \cdot CH : CH \cdot CH_2^-, Na^+ \xrightarrow{PhNHCH_3} CH_3 \cdot CH : CH \cdot CH_3$$

with simultaneous formation of 2 molecules of the sodium anilide.

The propagation of the ensuing polymerization was then described as an insertion of a monomer into the C–Na bond yielding another carbon-sodium bond – a description that differs only slightly from our present formulation. Not surprisingly, it was superfluous to postulate a termination step in this mechanism; the polymer still possessed the active C–Na bond. Indeed, polymerization was resumed when fresh monomer was added to the reactor[7].

An alternative interpretation of this polymerization was advocated by Schlenk and Bergmann[8]. In his early work reported in 1914[9] Schlenk described the addition of alkali metals to aromatic hydrocarbons leading to intensely colored solutions. The concepts of free radicals or radical-ions were unknown at that time, hence Schlenk referred to the adduct as a complex. He also showed that a similar reaction of 1,1-diphenyl ethylene resulted in its colored dimer which yielded 2,2,5,5-tetraphenyl adipic acid on carboxylation. When the concept of free radicals was established, Schlenk and Bergmann argued that the initially formed adduct is a free radical, e.g.

$$CH_2 : CH \cdot CH : CH_2 + Na \rightarrow NaCH_2 \cdot CH : CH \cdot CH_2 \cdot \ ,$$

and it initiates a radical type monomer addition. A similar reasoning accounted for the formation of the dimeric di-adduct of 1,1-diphenyl ethylene arising from radical dimerization. Interestingly, the latter interpretation comes close to our present description of this reaction – the recombination of radical-anions.

The success of radical theories of polymerization added credibility to the Schlenk and Bergmann mechanism and even caused some regression. For example, Schulz in 1938[10] and Bolland in 1941[11] were still upholding the idea of radical nature of diene polymerization induced by alkali metals. Nevertheless, slow progress was made. Abkin and Medvedev[12] demonstrated the long-life nature of the growing centers. They carried out polymerization of butadiene in an apparatus shown in Fig. 1. Metallic sodium was placed in vessel A and thereafter butadiene was distilled into it. After closing the upper stopcock, the progress of the reaction was observed by monitoring the pressure with the aid of an attached manometer. After awhile the monomer was distilled into vessel B and the connecting stopcock was closed. The constancy of pressure showed the absence of initiating or propagating species in B. Since radicals dimerize, they should disappear if kept in A for a sufficiently long time. However, when the monomer was back-distilled into A the polymerization resumed with its previous rate. The authors concluded, therefore, that the growing centers could not be radicals. The heterogeneous nature of the system nevertheless allows for an explanation invoking the presence of trapped radicals.

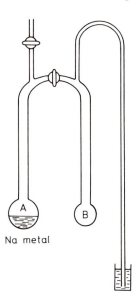

Na metal

Fig. 1. The apparatus of Abkin and Medvedev demonstrating the long-life nature of the growing species formed in polymerization of butadiene initiated by metallic sodium

In the following years several processes were attributed to anionic polymerization. Blomquist[13] reported anionic polymerization of nitroolefines initiated by hydroxyl ions. Beaman[14] recognized the anionic character of methacrylonitrile polymerization initiated in ether solution by Grignard reagents or by triphenylmethyl sodium. Robertson and Marion[15] reinvestigated the sodium initiated polymerization of butadiene in toluene and isolated oligomers having benzyl moiety as their end-groups. The characteristic red color developed in the course of that reaction implied the formation of sodium benzyl, presumably through proton transfer reaction from the solvent. Studies of homogeneous polymerization initiated by alkali metals or their amides in liquid ammonia, to be discussed later, left no doubt about their anionic character.

The final impetus for vigorous studies of anionic polymerization came in 1956 when two papers were published: the description of homogeneous electron-transfer initiated polymerization of styrene and isoprene yielding living polymers[16], and the discovery of 1,4-cis polymerization of isoprene initiated by metallic lithium in hydrocarbon solvents[17]. Since then the interest in this field has grown tremendously and its development has progressed in a truly exponential fashion.

I.2. A Brief Review of the Basic Steps of Polymerization

Ionic polymerization, like the well-known radical polymerization, is a chain process. As in other chain reactions, three basic steps should be distinguished in any addition type polymerization:

Initiation leads to the smallest entity which may reproduce the growing end-group by addition of a monomer.

Propagation is the process in which monomer molecules consecutively add to a growing center, regenerating it every time and producing an ever lengthening polymeric chain.

Finally, *termination* and *transfer* are the steps depriving the growing polymer of its capacity to spontaneously grow further. In a proper termination the ability of growth is lost irrevocably and the reaction may continue only through creation of new growing centers by some initiation step. In contrast, transfer reactions terminate the growth of a polymeric molecule simultaneously with the formation of a new growing center capable to continue the chain reaction.

While the growing end-groups in radical polymerizations are electrically neutral, they are charged in ionic polymerizations bearing positive charge in cationic and negative charge in anionic polymerization. There is, however, a variant of anionic polymerization in which the negative charge is transferred to a monomer, yielding an "activated monomer" that perpetuates the growth*.

Polymerizing systems are electrically neutral. Hence, some negatively charged ions are present in cationically polymerized systems, while cations neutralize the charges of anionically growing species. The participation of these counter-ions makes ionic polymerizations intrinsically more complex than the radical ones. Since the oppositely charged ions attract each other and strongly interact with the molecules surrounding them, a variety of species co-exist in the ionically polymerized medium. Free ions may exist in various solvation forms. The neutral associates, i.e. ion-pairs, are known to acquire a variety of shapes, e.g. tight or loose, depending on solvent, solvating agents added to the solution, temperature, etc. Further associations lead to triple-ions, quadrupoles etc., all potentially capable of influencing the course of ionic polymerization and affecting the character of each of its steps. The nature of these species and other topics related to their reactivities are discussed later.

I.3. Living polymers

Extensive studies of radical polymerization carried out in the period 1935–1950 firmly established the basic mechanism of the poly-addition[18]. Termination steps have been essential in accounting for the numerous observations, and their existence was unquestionable[19].

The imperative requirement of termination in radical polymerization arises from the nature of the interaction taking place between two radicals as they encounter each other. Coupling or disproportionation are the results, and in either case the interacting radicals are annihilated. However, the encounter between two ionically growing macromolecules does not annihilate them. Neither coupling nor disproportionation is feasible as two cations or two anions encounter each other. However, lack of a bimolecular termination involving two growing polymers does not exclude other kinds of termination or transfer, and the success of the conventional polymerization scheme created the impression that a non-terminated polymerization is highly improbable, if not impossible.

---

* The limited space prevents us from discussing this most ingeneous mechanism. Interested readers may consult Refs. 492–496

Polymerization schemes free of termination had been considered in earlier days. For example, the kinetic scheme of non-terminated polymerization was developed by Dostal and Mark[20] in 1935. Similarly, non-terminated, sodium-initiated polymerization of butadiene was visualized by Ziegler, and in fact the need of a termination step was not appreciated at that time. Later, several examples of non-terminated polymerization were considered by Flory[21], who also discussed some ramifications of such schemes.

However, it was not until 1956 that Szwarc and his associates[16] conclusively demonstrated the lack of termination in anionic polymerization of vinyl monomers in the absence of impurities. They proposed the term *"living polymers"* for those macromolecules which may spontaneously resume their growth whenever fresh monomer is supplied to the system. It should be stressed that living polymers, although not named in this way, were described earlier by Ziegler[7]. However, while the heterogeneous nature of these systems obscured some aspects of his work, and the ramifications were not emphasized strongly enough, Szwarc' studies were performed in homogeneous solution and all the important ramifications were clearly outlined. The original experiments were simple. The all glass apparatus depicted in Fig. 2 was used in their work. The device was thoroughly evacuated and then sealed off under vacuum. The initiator, a green solution of sodium naphthalene dissolved in 50 cm^3 of tetrahydrofuran, was admitted into reactor (A) by crushing a breakseal of the storage ampoule (B). Thereafter, 10 cm^3 of rigourously purified styrene, stored in ampoule (C), was slowly added into the reactor. As soon as the first drops of monomer came into contact with the initiator solution, its color changed abruptly from green to cherry-red, and then, within a second or two, the quantitative polymerization of styrene was completed. Nevertheless, although the reaction was over the cherry-red color persisted, suggesting that, perhaps, the reactive species were still present in solution.

By turning the reactor through 90°, the long side arm (S) was placed vertically and filled with the polymerized solution. An iron weight, enclosed in a glass envelope, was raised with the help of a magnet to a marked line and then allowed to fall freely. The time of fall, about 5 sec, provided an estimate of the viscosity of the investigated solution.

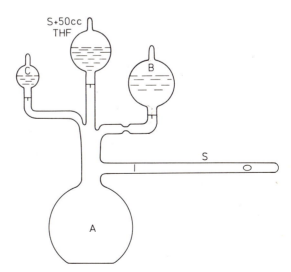

Fig. 2. The apparatus used by Szwarc et al. to demonstrate the living nature of anionically initiated polystyrene in a homogeneous THF solution

Thereafter, the reactor was restored to its original position, and a fresh batch of 10 cm^3 of styrene dissolved in 50 cm^3 of tetrahydrofuran was added to the polymerized mixture. Again, within seconds, the added monomer polymerized but although the color of the solution was not appreciably affected, its viscosity increased considerably. As shown by the falling weight method, the time of falling increased to about 50 s as compared to 5 s found previously. The higher viscosity of the solution could not be attributed to an increase in the concentration of the polymer; in fact, the latter remained unaltered, viz. 0.2 g/cm^3. This proved that the molecular weight of the polymer formed in the first stage of the experiment increased upon addition of fresh styrene. Hence, the original polymer, produced during the first polymerization, retained its ability to grow and became longer when a second batch of styrene was added.

To prove that *all* of the polystyrene formed in the first stage of the process was living, a modified experiment was performed. Instead of styrene, isoprene was added to the reactor in the second stage of the experiment. The total mass of the resulting polymer, determined after its precipitation, proved that the added isoprene had polymerized quantitatively. However, the isolated material, after being dissolved in toluene, could not be precipitated from such a solution by isooctane, whereas a 50 : 50 mixture of homopolystyrene and homopolyisoprene *could* be separated by this procedure, polystyrene being precipitated quantitatively. This result proved that no polystyrene was left after addition of isoprene, i.e. *all* of the original polymers were living and eventually produced block polymers of polystyrene-polyisoprene.

The characteristic features of living polymers are clearly revealed by these experiments. Living polymers may resume their growth whenever monomer is added to the system. They do not die but remain active and wait for the next prey. If the monomer added is different from the one previously used, a *block polymer* results. This, indeed, is the most versatile technique for synthesizing block polymers[22] (its details and applications will be discussed later).

Living polymers do not become infinitely long. Any system producing living polymers contains a finite amount of monomer. The system also contains some specified concentration of growing centers, or living polymers, and consequently all of the available monomer becomes partitioned among them. Hence, the number average degree of polymerization, $\overline{DP}_n$, is simply given by the ratio (total no. of moles of added monomer)/ (total no. of moles of living polymers or active end-groups).

The lack of natural death, i.e. of spontaneous termination, does not imply immortality either. The reactive end groups of living polymers may be annihilated by suitable reagents, a process known as "killing" of living polymers. It is desirable, indeed, to distinguish between a *"killing" reaction* and a *spontaneous termination*. The latter is governed by the law of probabilities and its course is set by the conditions existing *during* polymerization. On the other hand, the onset of the "killing" reaction is determined by the free choice of the experimenter, usually after completion of the polymerization. Moreover, he is free also to choose at will the reagent which converts active end groups into dead ones. For example, in the anionic polymerization of styrene propagated by carbanions, addition of water, or of any other proton-donating substance, converts the active $\sim\!\!\sim\!$ CH(Ph)$^-$ end groups into $>$C–H, whereas addition of carbon dioxide converts $>$C$^-$ into $>$C$\cdot$COO$^-$ and eventually into $>$C$\cdot$COOH. Similarly, a terminal hydroxyl group may be formed upon addition of ethylene oxide. It is important to realize that the polymers terminated by carboxyl or alkoxyl ions do not initiate polymerization of non-polar vinyl, vinylidene, or diene monomers. Therefore, in respect to their polymerization, they are dead. Nevertheless, they may initiate polymerization of some polar or ring-forming monomers, e.g. acrylonitrile, ethylene oxide etc., and hence they are still living as far as the polymerization of the latter monomers is concerned.

The "killing" technique is extremely useful since it allows us to introduce valuable functional end-groups into macromolecules, giving novel and interesting products. It is even more useful when applied to living polymers with two active end-groups since then *bifunctional* polymers are formed*.

Ideally, living polymers should retain their activities forever provided that they are not subjected to a "killing" action. However, in any real system this is not the case. Some feasible slow side reactions which annihilate the growing ends are unavoidable, and these set an upper limit to the durability of living polymers. Nevertheless, if the rates of these reactions are sufficiently slow to permit successful completion of a desired task, the system may correctly be classified as composed of truly living polymers.

I.4. Review of Some Living Polymer Systems

The earliest example of living polymers was furnished by Ziegler[7, 23)], who observed an increase in molecular weight of poly-butadiene initiated by metallic sodium on addition of fresh monomer to the system. However, it was not established whether *all* the polymers formed in this reaction resumed their growth or whether any residual or regenerated catalyst was left in the reactor. The latter could initiate then the formation of additional polymers. Due to these uncertainties, Ziegler's observations did not make the impact which otherwise could be expected, and did not induce further development in that field.

The living nature of ethylene oxide polymerization was recognized by Flory[21)], who conceived the ramifications of such a situation. However, the earlier studies of this reaction involved chain-transfer. The investigators added alcohol to the polymerizing solutions to solubilize the otherwise insoluble alkoxides and thus, the reaction

$$\text{Ɱ } CH_2O^-, Cat^+ + ROH \rightarrow \text{Ɱ } CH_2OH + RO^-, Cat^+$$

led to proton transfer and to relatively low molecular weight polymers. Subsequently, it was found that controlled hydrolysis of organic ferric salts yields catalysts capable of polymerizing oxiranes to high molecular weight products[37)], an approach adopted by Gee[38)] and by Osgan[39)]. Obviously, termination or transfer was greatly repressed in these systems.

Polymerization of ethylene oxide proceeding in the absence of alcohols was investigated by Figueruelo and Worsfold[51)], who chose hexamethyl phosphoric triamide as the solvent and sodium or potassium salts of $CH_3OC_2H_4OC_2H_4O^-$ as the initiator. The living character of this reaction was established.

Polymerization of propylene oxide initiated by t-butoxide salts in dimethyl sulphoxide showed the simple first order kinetics in the initiator as well as in the monomer[53)]. However, chain transfer to the solvent and E 2 elimination deprived this system of a living polymer character. The living character of ethylene oxide polymerization in te-

---

* Polymers endowed with two active end-groups are formed, e.g., by electron-transfer initiation.
See p. 43

trahydrofuran was demonstrated by Deffieux and Boileau[54] when the alkali counter-ions (K^+) were chelated by the (2,2,2)-kryptate.

Further improvement of oxiranes polymerization was reported by Inoue[24, 348]. Following his earlier studies of tetraphenyl-porphyrine-Al complexes, which revealad their usefulness as catalysts for propylene oxide polymerization and its co-polymerization with carbon dioxide[25], he modified the original complex by substituting $AlEt_2Cl$ for $AlEt_3$. The active species has the structure:

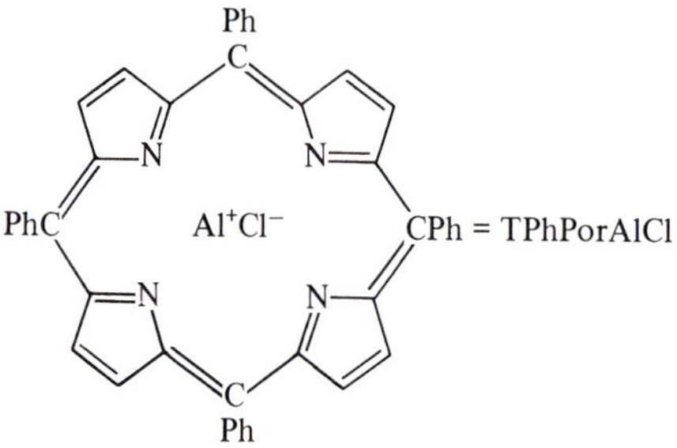

and it rapidly initiates polymerization of ethylene-, propylene-, or 1,2-butene-oxides. The reaction involves the insertion into the Al–Cl bond, i.e.

$$TPhPorAl-Cl \; + \; \underset{CH_2-CH_2}{\overset{O}{\triangle}} \; \longrightarrow \; TPhPorAlOCH_2CH_2Cl,$$

followed by the analogously proceeding propagation. Indeed, $ClCH_2CH_2(CH_3)OH$ was isolated when an equimolar mixture of the catalyst and propylene oxide were eventually hydrolyzed[26].

The resulting polymer showed a very narrow molecular weight distribution, and further evidence for a truly living character of this reaction was provided by the preparation of di- and tri-block polymers of the oxiranes with exclusion of the homo-polymers.

The nature of the catalysts resulting from the controlled hydrolysis of organo-metallic compounds was extensively studied by Teyssié. Following the previous observations of Tsuruta[40] and Vandenberg[41], who prepared useful catalysts by hydrolysis of zinc and aluminum alkyls, he investigated two-step condensation between metal acetates and alkoxides. This procedure led to a well-defined catalyst of the following structure[42]:

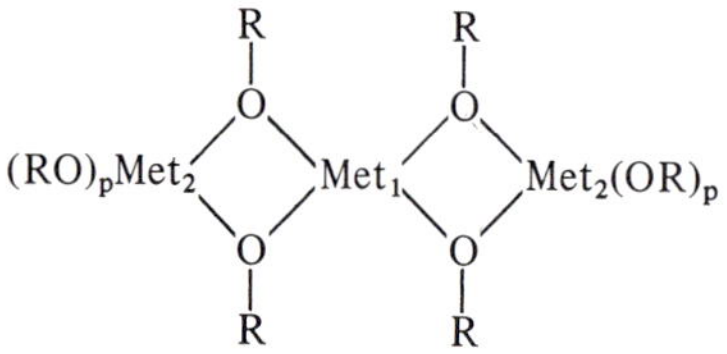

where Met_1 stands for Zn^{II}, Cr^{II}, Mo^{II}, Fe^{II}, or Mn^{II}, while Met_2 represents Al^{III} or Ti^{IV}, and the subscript p is 2 for Al^{III} and 3 for Ti^{IV}. These mixed bimetallic alkoxides are aggregated; e.g. in benzene or cyclohexane they form octamers, although they are monomeric in butanol[43]. The aggregation probably explains their remarkable solubility in hydrocarbons arising from a compact oxide structure surrounded by a lipophilic layer of the alkoxide groups.

The mixed bimetallic alkoxides rank among the best catalysts of ring-opening poly-merization of oxiranes, thiiranes, and lactones. The one formed from Al(OBu)$_3$ and zinc acetate seems to be superior. Kinetic and structural data suggest a coordinative anionic mechanism of initiation and propagation. The monomer is inserted into an Al–OR bond simultaneously with its opening, like in the proposed flip-flop mechanism of Vanden-berg[44]. The reactivity of the catalyst and its selectivity in co-polymerization could be greatly altered by varying solvents and the nature of the R groups.

Polymerization of ε-caprolactone initiated by the above catalyst shows all the charac-teristic features of living polymer system[45]. The degree of polymerization increases with conversion and the reaction is resumed on addition of fresh monomer. If carried out in butanol, which dissociates the aggregates, the product has a narrow molecular weight distribution with the M_w/M_n ratio approaching 1.05[46, 47]. Under these conditions all four terminal butoxy groups participate in the reaction and hence the number average degree of polymerization is given by the ratio of moles of the polymerized monomer to 4 Zn. On hydrolysis of the polymer-catalyst complex, one obtains the hydroxyl terminated poly-mers possessing a $-CH_2 \cdot OBu$ group on their other end – a result expected on the basis of the proposed mechanism.

Finally, the living character of that polymerization permits the preparation of block polymers. Consecutive addition of two different lactones, e.g. caprolactone and β-pro-piolactone, readily led to the expected two-block polymers. The virtually quantitative conversion and the exclusion of homopolymers were attained under complete dissocia-tion of the aggregates. Otherwise, as the second monomer is added, some inactive, sterically hindered OR groups may become available for the initiation. This would yield some homo-polymers. Alternatively, addition of caprolactone followed by propylene oxide resulted in preparation of the respective two-block polymers.

Anionic polymerization of thiiranes, e.g. 2-methyl-2-ethyl thiirane[27], initiated by carbazyl or fluorenyl salts in tetrahydrofuran, also yields living polymers, see p. 152.

Numerous vinyl, vinylidene and diene monomers yield living polymers when anioni-cally polymerized under proper conditions. These reactions were extensively studied and specific systems revealing many details of their conduct will be discussed later.

The anionic mode of polymerization is not the only one leading to formation of living polymers. Cationic polymerization of many cyclic ethers, sulphides, amines etc., also yields living polymers, provided that the counter-ions are judiciously chosen. The first example of living, cationically growing polymer was provided by the simultaneous but independent studies of three groups[29–31] who investigated the polymerization of te-trahydrofuran. This subject has been comprehensively reviewed by Penczek and his associates[28].

Some coordination polymerizations also exhibit the character of a living polymeriza-tion process. For example, Natta[32] described a heterogeneous catalyst yielding block polymers when the initially present monomer was replaced by another one. In that system, the lifetime of a growing polymer exceeds one-half hour, i.e. the termination caused by the hydride transfer to the catalytic center was very slow. Similar results were reported by Bier[33] and by Kontos et al.[34].

The most spectacular coordination system free of termination and transfer was described recently by Doi et al.[35]. The homogeneous vanadium acetyl acetonate com-plex with AlEt$_2$Cl rapidly initiated polymerization of propylene at $-60\,°C$. The molecular weight of the product increased with conversion and the ratio M_w/M_n was only slightly greater than 1.0. These facts confirm the rapidity of initiation and the lack of termination or chain transfer. Interestingly, the above catalyst was described previously by Natta[36], who utilized it for synthesis of syndiotactic poly-propylene.

Finally, two more examples of living polymer systems should be mentioned. Polymerization of diazomethane initiated by boron trifluoride gives a stable compound[48]

$$\sim\!\!\sim CH_2 \cdot BF_2$$

which reacts with diazomethane and perpetuates further growth of the polymer, i.e.

$$\sim\!\!\sim CH_2BF_2 + CH_2N_2 \rightarrow \sim\!\!\sim CH_2CH_2BF_2 + N_2 \; .$$

Polymerization of chloral initiated by organo-Sn compounds yields[49]

$$\sim\!\!\sim \underset{CCl_3}{\overset{|}{CH}}\!-\!O\!-\!Sn(CH_3)_3$$

The latter reacts with chloral by insertion into the Sn–O bonds,

$$\sim\!\!\sim \underset{CCl_3}{\overset{|}{CH}}\!-\!O\!-\!Sn(CH_3)_3 \;+\; \underset{CCl_3}{\overset{|}{CHO}} \longrightarrow \sim\!\!\sim \underset{CCl_3}{\overset{|}{CH}}\!\cdot\! O\!-\!\underset{CCl_3}{\overset{|}{CH}}\!-\!O\!-\!Sn(CH_3)_3$$

hence the growth continues as chloral is supplied to the reactor.

I.5. Living and Dormant Polymers

In some systems most of the polymers potentially capable of growing are inert and do not contribute to propagation. The polymerization involves a small fraction of active polymers remaining in dynamic equilibrium with the inert ones. We refer to the latter as the *dormant polymers* in equilibrium with the living ones. A few examples illustrate this behavior:

(a) In anionic polymerization most of the polymers are often present in the form of tight ion-pairs that virtually do not contribute to the propagation carried out by free ions. The latter are then the living polymers in equilibrium with the dormant ion-pairs.

(b) Lithium salts of living polystyrene, polybutadiene etc., are present in hydrocarbon solvents as inactive dormant aggregates, e.g. dimers. The propagation arises from the presence of a minute fraction of active unaggregated polymers, the living ones. An even more complex situation is encountered in the presence of lithium alkoxides. According to Roovers and Bywater[50], the following equilibrium is established:

$$(\sim\!\!\sim styrene^-, Li^+)_2 \quad\overset{LiOR}{\rightleftharpoons}\quad (\sim\!\!\sim styrene^-, Li^+, LiOR)_2 \qquad - \text{ dormant}$$

$$\Updownarrow \qquad\qquad\qquad\qquad \Updownarrow$$

$$2(\sim\!\!\sim styrene^-, Li^+) \qquad 2 (\sim\!\!\sim styrene^-, Li^+, OR) \qquad - \text{ living}$$

$$\quad\text{more reactive} \qquad\qquad\quad \text{less reactive}$$

(c) In cationic polymerization of tetrahydrofuran with $CF_3SO_3^-$ counterions, an equilibrium is established between the inert macro-esters,

$\sim\!\!\sim (CH_2)_4O \cdot SO_2CF_3$,

and the reactive macro-ions

$\sim\!\!\sim (CH_2)_4\overset{+}{O}\!\!\diagup\!\!\bigcirc$, $CF_3SO_3^-$

Similarly, in cationic polymerization of oxazoline initiated by methyl-iodide, the inert polymer terminated by the CH_2I group is in equilibrium with its reactive ionized form.

The presence of dormant polymers does not affect the character of propagation perpetuated by the active polymers. However, the concentration of the active end-groups depends on the equilibrium established between the active and dormant groups, and consequently the observed rate of propagation is affected accordingly. Whenever the lifetime of the dormant polymer is short compared with the time between two consecutive monomer additions, this phenomenon does not affect molecular weight distribution, otherwise it leads to its broadening.

I.6. Transformation of Active End-Groups of Living Polymers

Living polymers possess active, growth sustaining end-groups on one or both of their ends. In anionic polymerization of vinyl, vinylidene or diene monomers, these are carbanions which easily are converted into other functional groups. Their conversion into carboxylic or hydroxylic groups was mentioned earlier; and introduction of terminal halogens, nitriles, aldehydes, mercapto, etc., groups was reported by Uraneck et al.[452]. Although in principle the procedure is simple, a successful synthesis is difficult. In order to produce polymeric molecules, *each* being endowed with the desired functional end-group, several conditions have to be fulfilled. Any spontaneous or accidental termination has to be avoided, the chosen reagent has to react quantitatively with living polymers and no side reactions should take place, the product must not react with the not yet converted living polymers, etc.

Introduction of some sophisticated functional groups was described by Rempp and his colleagues, e.g. acyl lactams, isocyanates[453], vinyl silanes[454], phosphoric esters[455] etc. Attachment of dyes or fluorescing groups is of considerable interest in various academic studies and progress in this field has been reported. However, the most interesting is the introduction of polymerizable end-groups – exemplified by the preparation of macromers described by Milkovich[456]. These are monomers with attached long chains, and since living polymer technique is used for their preparation, their length is easily controlled and their molecular weight distribution could be very narrow. Macromers could be co-polymerized with other monomers by a variety of techniques, not necessarily ionic. This leads to formation of branched-chain polymers having a controllable number of well characterized branches. The benefits arising from their presence are illustrated by several examples reported by Milkovich[456].

Another interesting and potentially most promising development is anticipated from studies of Richard[457] as well as of other investigators. They explored the idea of converting anionically growing end-groups into others capable of initiating further growth by some alternative mechanism. In essence, the following transformations could be visualized: anionic group converted into free radical or cationically propagating end-groups, a free radical into anion or cation sustaining further growth, and cationically propagating group into the other two. Moreover, there are ways to convert anions into metalloorganic complexes like the Ziegler-Natta kind. In this way, monomers that could not be polymerized anionically could be used in propagation yielding interesting block polymers.

Some routes leading to transformation of carbanions into cations were described[458]. One of the methods involves capping carbanions with bromine and conversion via Grignard into reactive benzyl bromides that could be oxidized by silver salts. Alternatively, phosgene could be used as the reagent[459]. Conversion of cations into anions is exemplified by the reaction of living polytetrahydrofuran with lithium salt of cinamyl alcohol,

$$\text{ⱵⱵⱵ } \overset{+}{O}\!\!\diagdown \qquad + \ \ LiOCH_2 \cdot CH{:}CH_2(Ph) \ \longrightarrow \ \ \text{ⱵⱵⱵ } O(CH_2)_4 \cdot OCH_2 \cdot CH{=}CH(Ph),$$

followed by initiation of polymerization with butyl lithium[460]*. Conversion of anion to free radical may be achieved by reacting a living polymer, e.g. living polystyrene, with triethyl lead chloride. The thermal decomposition of the resulting adduct, ⱵⱵⱵ $CH(Ph) \cdot PbEt_3$, yields a radical. Mercury diethyl might be used instead of the lead compound. Formation of a Ziegler-Natta type of catalyst is achieved by reacting a living polymer with, e.g. $AlCl_3$ and forming

$$Al(M \text{ ⱵⱵⱵ})_3 \quad or \quad AlCl_2(M \text{ ⱵⱵⱵ})$$

These could be used with transition metals to produce reactive centers[461].

Finally, some spontaneous reactions convert the living end into a dead one. The trivial example of protonation was mentioned previously. Some solvents might act as proton donors, e.g.

$$\text{ⱵⱵⱵ } \overline{C}H(Ph), Li^+ + PhCH_3 \rightarrow \text{ⱵⱵⱵ } CH_2(Ph) + PhCH_2Li \ ,$$

or deactivate the growing carbanion in some alternative way, e.g.

$$\text{ⱵⱵⱵ } \overline{C}H(Ph) + THF \rightarrow \text{ⱵⱵⱵ } CH(Ph)CH_2CH_2\overline{O} + C_2H_4 \ .$$

Spontaneous reactions yielding highly colored but inactive end-groups are particularly troublesome. Only one example of such a reaction is fully understood, namely the conversion of living polystyrene into 1,3-diphenyl allyl anion with simultaneous formation of sodium hydride[462].

In conclusion, the chemistry of growing end-groups is a developing field, and new methods of the extension of polymer reactions are expected.

---

* Preparation of block polymers through this approach was reported by Schué and Richards[476]

II. Thermodynamics of Polymerization

II.1. Thermodynamics of Propagation

The basic approach to thermodynamics of propagation was outlined by Dainton and Ivin[179] and the subject was comprehensively reviewed by Ivin[189]. This step of addition polymerization is described by the scheme

$$\text{\Large\textasciitilde} \text{MM}^* + \text{M} \underset{k_d}{\overset{k_p}{\rightleftharpoons}} \underset{n+1\text{-mer}}{\text{\Large\textasciitilde} \text{MMM}^*} , \quad K_p ,$$

n-mer

where M denotes a monomer, the asterisk signifies the ability of the last unit of the respective polymer to perpetuate its propagation, and K_p is the equilibrium constant of propagation. Like any other elementary reaction, propagation is reversible, k_p and k_d denoting the respective propagation and depropagation rate constants.

For a sufficiently long polymeric chain, K_p is independent of its degree of polymerization since the above reaction converts a monomer into a monomeric segment placed in the midst of a long polymeric chain. Hence, this process may be symbolically described as

$$M_f \rightleftharpoons M_s$$

where M_f is a monomer in a solution and M_s is a monomeric segment of a long polymer. It follows from the above description that K_p is unaffected by the mechanism of propagation, i.e. by the character of the propagating unit. However, its value depends not only on the nature of the monomer but also on the nature and composition of a solution in which the reaction proceeds. A change of environment affects the free energies of the monomer and monomeric segments of the polymer; hence, it influences the value of K_p. Quantitative treatment of such effects was reported first by Bywater[180], who dealt with this problem in terms of the Flory-Huggins theory of polymer solution adopted to ternary systems.

In conventional addition polymerizations the growing chains are formed by some initiation processes and destroyed by some virtually irreversible terminations. The conversion of monomer into polymer eventually could be quantitative, provided that the initiation continues throughout the process. In the absence of termination or chain transfer the growing polymers remain living and then the polymerizing system ultimately has to attain a state in which the living polymers are in equilibrium with their monomer. The equilibrium concentration of the monomer, M_e, provides valuable information leading to the determination of the appropriate K_p. To clarify this point, let us consider the equilibria

$$IM^* + M \rightleftharpoons IMM^* \qquad\qquad K_1$$

$$IMM^* + M \rightleftharpoons I(M)_2M^* \qquad\qquad K_2$$

. . . .

$$I(M)_iM^* + M \rightleftharpoons I(M)_{i+1}M^* \qquad\qquad K_{i+1}$$

. . . .

Although K_1 need not be equal to K_2, and neither need K_2 be equal to K_3, it is obvious that for larger i's $K_i = K_{i+1} = \cdots = K_p$. Without loss of generality we may assume that $K_1 = K_2 \cdots = K_i = K_p$, since this assumption does not alter the essence of the forthcoming conclusions. Furthermore, let us restrict our consideration to ideal systems when activities are equal to concentrations. Under those assumptions the following relations are valid:

$$M_0 - M_e = (IM^*)_e/(1 - K_pM_e)^2$$

and

$$I_0 = (IM^*)_e/(1 - K_pM_e) \ ,$$

where M_0 and I_0 denote the initial concentrations of the monomer and of the quantitatively utilized initiator, respectively, and $(IM^*)_e$ is the equilibrium concentration of the first product of initiation. Hence,

$$M_e = \{K_pM_0 + 1 - \sqrt{(K_pM_0 - 1)^2 + 4K_pI_0}\}/2K_p \ .$$

The equilibrium concentration of the monomer depends, therefore, not only on K_p but also on the initial concentrations of the monomer and initiator. Alternatively, M_e is given by the relation

$$M_e = (1 - 1/\overline{DP}_n)/K_p$$

where $\overline{DP}_n$, the number average degree of polymerization, is given by $(M_0 - M_e)/I_0$. Either form leads to a simplified relation whenever polymers of a high degree of polymerization are formed, i.e. for $M_0 \gg I_0$ and $\overline{DP}_n \gg 1$. In such cases

$$M_e \approx 1/K_p \ .^*$$

In real systems additional problems arise. The larger M_0 the higher the concentration of the polymer in the ultimately formed solution. This, in turn, affects the activity coefficient of the monomer and therefore the value of M_e. Similarly, in ionic polymerization a high concentration of the initiator leads to a high concentration of ionic species – another factor substantially affecting the monomer's activity coefficient. Examples of such phenomena are given later.

---

* Note that at $I_0 \to 0$ the monomer equilibrium concentration, M_e, is given by $1/K_p$ for $K_p > 1$ but by M_0 for $K_p < 1$, the system exhibits a second-order transition at extremely low concentration of initiator – a discontinuity of dM_e/dT at temperature T for which $K_p = 1$. See Refs. 488–490

The first application of living polymers in thermodynamic studies was reported by Worsfold and Bywater[181] and by McCormick[182]. In both studies anionic polymerization of α-methyl styrene, initiated in tetrahydrofuran by electron-transfer, was investigated over temperatures ranging from $0\,°C$ to $-40\,°C$. The results are shown graphically in Fig. 3, and lead to heat of polymerization of -8 kcal/mol, entropy of polymerization in solution of ~ 29 eu, and to about 1 M equilibrium concentration of α-methyl styrene at ambient temperature.

The effect of increasing concentration of polymer was noted in the work of Vrancken et al.[183] as shown in Fig. 4. An increase of the *total* concentration of poly-α-methyl styrene in tetrahydrofuran, whether caused by the addition of increasing amounts of monomer or by the addition of a dead polymer, decreases the equilibrium concentration of α-methyl styrene. An extension of these studies[184] led to a relation between monomer equilibrium concentration and the volume fraction of the polymer. For the system α-methyl styrene-tetrahydrofuran solvent,

$$M_e = M_{e0} - B\phi_p ,$$

where M_{e0} is the limiting value of M_e, as the volume fraction of the polymer tends to 0, while B is a constant that increases with temperature[184].

Anionic polymerization was utilized again in studies of thermodynamics of styrene propagation[185]. The equilibrium concentration of that monomer is exceedingly low at ambient temperature, and hence the experimentation required elevated temperatures. However, living polystyrene in THF is rapidly destroyed at those temperatures. To avoid these difficulties, living polystyrene formed by BuLi initation in cyclohexane or benzene was used in the studies. The results are presented in Fig. 5 and in Table 1. The effect of the solvent's nature on M_e is revealed by these data.

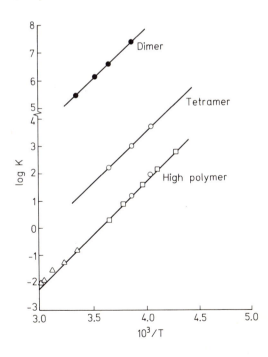

Fig. 3. Equilibrium concentration, M_e, of α-methyl styrene as a function of temperature in THF. $M_e = 1/K_\infty$. $\square$ – Results of Worsfold and Bywater; $\triangle$ – Results of McCormick. $\bullet$ – Equilibrium constant for K^+, $^-aa^-$, $K^+ + a \rightleftharpoons K^+$, $^-aaa^-$, K^+; $\bigcirc$ – Equilibrium constant for Na^+, $^-aaaa^-$, $Na^+ + a \rightleftharpoons Na^+$, $^-aaaaa^-$, Na^+; a denotes α-methyl styrene; $^-aa^-$ $^-aaaa^-$ denote its dimeric and tetrameric dianions

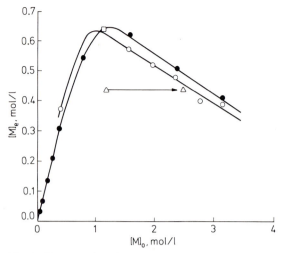

Fig. 4. Concentration of α-methyl styrene, M_e, in equilibrium with its living oligomer or polymer as a function of added amount of the monomer, M_0, at constant temperature in THF solution. M_e increases with M_0 since the initially present dimeric dianions are converted into oligomers and eventually polymers. M_e should reach a plateau as high molecular weight polymer is formed. However, increasing concentration of the polymer increases the activity coefficient of the monomer, causing a decrease of M_e. The point denoted by a △ results from the addition of a dead polymer to a solution corresponding to a point □. Hence, M_e decreased

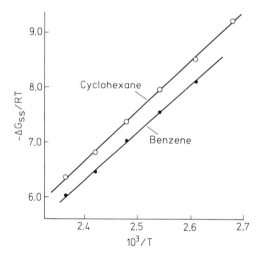

Fig. 5. Equilibrium between dilute solution of lithium polystyrene and styrene at different temperatures. $-\Delta G_{ss}/RT = \ell nK_\infty = -\ell nM_e$. Solvents cyclohexane (*open points*) and benzene (*filled points*)

Anionically formed polymers are not the only kind used in such studies. Cationically prepared living polymers were extensively utilized; in fact, the attainment of living polymer-monomer equilibrium provided the first evidence for the living character of cationically polymerized tetrahydrofuran[188]. In that polymerization the cyclic oxonium ions are the propagating species,

〜 $OCH_2CH_2CH_2CH_2\overset{+}{O}$⬠

Table 1. Equilibrium concentration of styrene in equilibrium with living polystyrene

T°C	$M_e \times 10^5/M$ in Benzene	$M_e \times 10^5/M$ in Cyclohexane
100	–	3.97
110	12.0	7.79
120	20.7	13.6
130	34.2	23.8
140	59.0	41.6
150	90.8	65.0

A nucleophilic attack by the oxygen of another THF molecule on the carbon adjacent to O^+ results in opening of the ring, formation of a new C–O bond, and a new cyclic oxonium ion. The equilibria between living poly-tetrahydrofuran and its monomer were attained from both directions, by raising the temperature and causing depropagation, or by lowering it and inducing propagation[188a]. The equilibrium concentration of THF at ambient temperature is $\approx 3\,M$.

The influence of the polymer concentration upon $[THF]_e$ was investigated by Ivin and Leonard[193] and the relation

$$[THF]_e = [THF]_{e,0} + B\phi_p$$

was again found to be valid. Polymerization carried out in a variety of solvents revealed their effect upon the equilibrium. The results obtained at a constant temperature in CCl_4, CH_2Cl_2, CH_3NO_2 and C_6H_6 are presented in graphic form in Fig. 6, giving $[THF]_e$ as a function of its initial concentration, $[THF]_0$[191]. Of course, as required, the lines converge for $[THF]_0 \approx 12\,M$ corresponding to bulk THF.

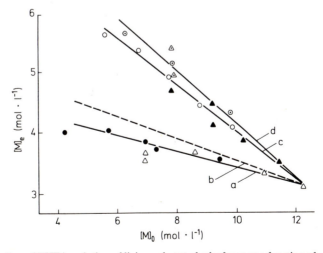

Fig. 6. Equilibrium concentration of THF in solution of living poly-tetrahydrofuran as a function of initial concentration of THF, T = 25 °C. Solvents: (a) CCl_4, (b) C_6H_6, (c) CH_2Cl_2, (d) CH_3NO_2

The apparent discrepancy between the results of Saegusa[497] and of Penczek[191] reflects the effect of the concentration of ionic species upon the equilibrium. Different concentrations of initiator were used by these workers; and the higher its concentration, the higher the concentration of oxonium ions in the system. This, in turn, reduces the activity coefficient of the ether leading to higher $[THF]_e$.

A general treatment of the dependence of M_e on the nature of solvent and the volume fraction of polymer was outlined by Ivin and Leonard[184]. Adopting the Flory-Huggins theory of polymer solutions to ternary systems, they derived the relation

$$\Delta G_{\ell,c}/RT = \ell n \phi_m + 1 + \phi_s(\chi_{ms} - \chi_{sp}V_m/V_s) + \chi_{mp}(\phi_p - \phi_s) \,,$$

where $\Delta G_{\ell,c}$ is the free energy of conversion of 1 mole of liquid monomer into 1 base-mole of amorphous polymer; the ϕ's are the volume fractions of the monomer, solvent, and the polymer in the equilibrated mixture; the χ's are the monomer-solvent, polymer-solvent, and monomer-polymer interaction parameters; and V_m and V_s are the molar volumes of the monomer and solvent, respectively. This relation is more general than that previously deduced by Bywater[180], who implicitly assumed $V_m = V_s$. In the absence of solvent, i.e. for equilibrium established in a bulk monomer which is a solvent for its own polymer, the above relation is reduced to

$$\Delta G_{\ell,c}/RT = \ell n \phi_m + 1 + \chi_{mp}(\phi_p - \phi_s) \,.$$

The validity of this relation was checked experimentally[184, 190].

II.2. The Significance of the Molecular Weight of Living Polymers

The simplified relation $M_e = 1/K_p$ is applicable to high-molecular weight polymers only; it fails for oligomers. Let it be stressed that the concentration of living polymer molecules *does not* affect the equilibrium concentration of a monomer*, provided that their volume fraction is very low. The previously discussed effects arise from a change of environment which affects the activity coefficient of the monomer. However, the size of living polymers becomes a factor when the polymers are short. To appreciate this point, let us consider the following argument.

At equilibrium the rates of propagation and depropagation exactly balance each other. We may assume, without loss of generality, that the rate constants k_p and k_d are independent of the degree of polymerization. Hence,

$$k_p[M]_e \sum [\text{living } i\text{-mers}] = k_d \sum{}' [\text{living } i\text{-mers}]$$

The first summation includes all the living polymers capable of growing, whereas the second summation includes all the polymers capable of degrading. Of course, there has to be a lowest living polymer, e.g. IM* or *M–M*, which can grow but not degrade. Whenever high-molecular weight polymers are formed, the approximation

$$\sum [\text{living } i\text{-mers}] = \sum{}' [\text{living } i\text{-mers}]$$

---

* See, however, the footnote on p. 16

is acceptable since the mole fraction of the lowest polymers, IM* or *MM*, is exceedingly low. However, this is not the case for low molecular weight oligomers. The respective equilibrium monomer concentration is then lower than that obtained for high-molecular weight polymers.

An example is provided by the study[192] of the addition of α-methyl styrene, α, to its dimeric dianion,

$$K^+, \overline{C}(CH_3)(Ph) \cdot CH_2 \cdot CH_2 \cdot \overline{C}(CH_3)(Ph), K^+ = K^+, \, {}^-\alpha\alpha^-, K^+ \,.$$

The investigated reaction is represented by the scheme,

$$K^+, \, {}^-\alpha\alpha^-, K^+ + \alpha \underset{k_b}{\overset{k_f}{\rightleftharpoons}} K^+, \, {}^-\alpha\alpha\alpha^-, K^+, \quad K = k_f/k_b \,,$$

and its kinetics was investigated in THF solution by the stirred-flow-reactor technique leading to the relation,

$$1/\tau = 2k_f[\alpha]_s\{[K^+, \, {}^-\alpha\alpha^-, K^+]_0/([\alpha]_0 - [\alpha]_s) - 1\} - k_b \,.$$

Here, τ is the residence time of the reactants, and the subscripts 0 and s refer to their concentrations in the feed and in the reactor, respectively. It follows that a plot of $1/\tau$ vs. $[\alpha]_s\{[K^+, \, {}^-\alpha\alpha^-, K^+]_0/([\alpha]_0 - [\alpha]_s) - 1\}$ is linear, with a slope $2k_f$ and the intercept $- k_b$. Such a plot is shown in Fig. 7, and the results obtained for different compositions of the feed are collected in Table 2.

The average value of that equilibrium constant, namely 331 M^{-1}, should be compared with that governing the equilibrium with a high-molecular weight poly-α-methyl styrene,

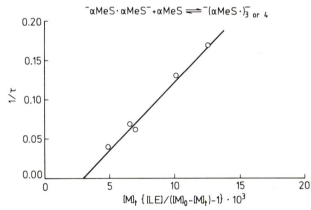

Fig. 7. Equilibrium and kinetics of addition of α-methyl styrene, α, to its dimeric dianion, K^+, ${}^-\alpha\alpha^-$, K^+, in THF studied by stirred-flow technique. T = 25 °C.

$$K^+, \, {}^-\alpha\alpha^-, K^+ + \underset{k_b}{\overset{k_f}{\rightleftharpoons}} K^+, \, {}^-\alpha\alpha\alpha^-, K^+$$

τ – residence time in the stirred reactor, $[K^+, \, {}^-\alpha\alpha^-, K^+]_0$ – initial concentration of the dimer, $[\alpha]_0$ and $[\alpha]_s$ – the concentrations of the monomer in the feed and in the reactor, respectively
$1/\tau = 2k_f[\alpha]_s\{[K^+, \, {}^-\alpha\alpha^-, K^+]_0/([\alpha]_0 - [\alpha]_s) - 1\} - k_b$

Table 2. The rates and equilibrium constants for the reversible reaction. K^+, $^-\alpha\alpha^-$, $K^+ + \alpha \rightleftharpoons K^+$, $^-\alpha\alpha\alpha^-$, K^+

$10^3 [K^+, {}^-\alpha\alpha^-, K^+]_0/M$	$10^3 [\alpha]_0/M$	$k_f Ms$	$k_b s$	K.M.
3.0	3.0	17.2	0.050	346
4.4	2.9	17.3	0.052	333
4.5	1.1	16.7	0.049	339
6.5	2.9	18.9	0.062	304
14.0	3.0	15.4	0.046	333
Average		17.1	0.052	331

vis $\sim 1\ M^{-1}$. This enormous increase is attributed to a difference in steric strain associated with an addition of the monomer. The addition of a molecule of α-methyl styrene to its high-molecular weight polymer, in fact even to a tetramer derived from $^-\alpha\alpha^-$ or a dimer $I\alpha\alpha^-$, involves a steric strain arising from the necessity of squeezing one more unit between the bulky neighbors. In contrast, the addition to a dimeric dianion (remember its structure) is hindered by a resistance from only one bulky neighbor.

An attempt to determine the sequential equilibrium constants, K_1, K_2, etc., was reported by Vrancken et al.[183]. They initiated polymerization of α-methyl styrene by a stepwise addition of the monomer to a dilute solution of its dimeric dianions in tetrahydrofuran. After each addition the concentration of the residual, equilibrated monomer was determined, and its value was plotted vs. the total concentration of the supplied monomer. The resulting curve is shown in Fig. 4, and its shape allows, in principle, the determination of K_1, K_2 and $K_p = K_3 = K_4$, etc. Although the approach is sound, this method gives only a reasonable estimate of K_1 because the deviation of K_2 from the K_p's value is too small to be determined. The maximum seen in Fig. 4 arises from an increase in the activity coefficient of the monomer caused by increasing volume fraction of the polymer – an effect discussed earlier.

II.3. The Temperature Dependence of M_e

The temperature dependence of M_e provides the information needed for calculating the heat, ΔH_p, and the entropy, ΔS_p, of polymerization. The heat of polymerization was obtained in the earlier studies from the slope of a plot of $\ell n\ M_e$ vs. $1/T$. This procedure is not quite correct. The volume fraction of the polymer increases with decreasing temperature; and in the bulk polymerization, it approaches unity at sufficiently low temperatures. This, as has been pointed out earlier, affects the activity coefficient of the monomer. The proper procedure calls for a plot of $\Delta G_{\ell,c}/RT$, calculated from the previously discussed relation, vs. $1/T$. Such plots referring to the THF system in the bulk and in benzene solution[190] are shown in Fig. 8. The two lines should overlap; failure to do so arises from uncertainties in the values of the interaction parameters χ needed in the calculations. Nevertheless, it is gratifying to find closely similar values of ΔH_p from both slopes, namely -3.0 and -3.3 kcal/mol, respectively.

Most polymerizations are exothermic and decrease the entropy of polymerizing systems; hence, M_e increases with rising temperature. The highest possible value of M_e is

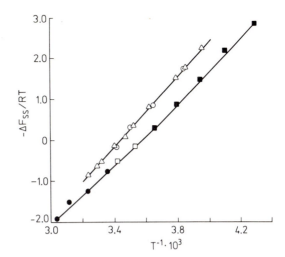

Fig. 8. Plot of $-\Delta G_{ec}/RT$ vs. $1/T$. Monomer THF. For benzene solution $\triangle$ and for bulk THF $\bullet$

that corresponding to the concentration of a monomer in the bulk, and the temperature at which M_e of bulk polymerization reaches that value is known as the ceiling temperature, T_c, of that monomer. Obviously, its polymerization to high-molecular weight polymer is impossible above T_c.

Some polymerizations proceed endothermally and increase the entropy of polymerizing systems. Polymerization of sulphur, S_8, into long chains of plastic sulphur is the best known example of such a case[194]. In such systems M_e increases with decreasing temperature and below a critical temperature T_f, known as a floor temperature, polymerization is thermodynamically forbidden. Obviously, T_c or T_f, are given by the relation

$$T_c = \Delta H_p/\Delta S_p$$

where ΔH_p and ΔS_p refer to the heat and entropy of conversion of a bulk monomer into its amorphous polymer.

The notion of the critical temperatures, T_c and T_f, might imply a sharp restriction for conversion of a monomer into a polymer or oligomer. This is not the case. For example, in an exothermic reaction no *high*-molecular weight polymer could be formed above the ceiling temperature*; however, this restriction does not apply to low molecular weight oligomers. At any initial concentration, M_0, of a monomer and at concentration I of the quantitatively utilized initiator, the number average degree of polymerization of oligomers or polymers formed at equilibrium changes with temperature, as shown in Fig. 9[197]. The temperature corresponding to the inflection point of the curve shown in that figure depends on the ratio M_0/I, and the steepness of the curves at that point increases with increasing heat of polymerization. Similar sigmoidal curves are obtained for the fraction of polymerized monomer as a function of temperature as shown in Fig. 10. These results are deduced from the general treatment outlined in Sect. II.1 giving $M_e = (1 - 1/\overline{DP_n})/K_p$. The dependence of the number average degree of polymerization on M_0 at constant temperature and for constant I is shown by Fig. 11.

---

* For a finite I_0. However, for $I_0 \rightarrow 0$ a high-molecular polymer could be formed above T_c. See Ref. 488

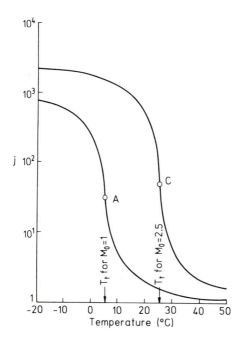

Fig. 9. Plot of the number average degree of polymerization, j, of a living oligomer or polymer vs. temperature at constant total amount of supplied monomer, M_0, and a constant I

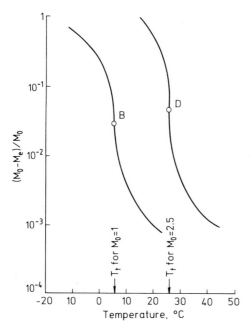

Fig. 10. The fraction of monomer converted into a living oligomer or polymer, $(M_0-M_e)/M_0$, as a function of temperature at constant I

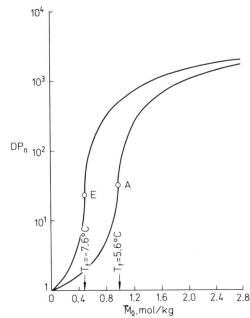

Fig. 11. The number average degree of polymerization as a function of total amount of supplied monomer, M_0, at constant temperature and for a constant I

II.4. Molecular Weight Distribution in Equilibrated Living Polymer Systems

Living polymers resulting from an instantaneously initiated but non-terminated polymerization, have a nearly Poisson molecular weight distribution, provided that $M_0 \gg M_e$. The polymerization seems to cease as the concentration of the residual monomer attains its equilibrium value – no further conversion of the monomer into polymer could be detected at that stage of the reaction. Nevertheless, the system is not yet in its ultimate equilibrium state.

In a truely equilibrated living polymer system, the molecular weight distribution is of a Flory type, i.e. $[P^*_{i+1}]/[P^*_i] = K_p M_e = \text{const}^*$. Hence, after the pseudo-equilibrium state has been attained, the continuous depropagation and propagation reactions gradually change the molecular weight distribution from Poisson to Flory type.

This is a peculiar process. It does not change the number of polymeric molecules, virtually does not affect their total mass, and hence leaves their number average degree of polymerization virtually constant. The concentration of the residual monomer hardly changes; in fact, it decreases from $1/K_p$ to $(1 - 1/\overline{DP}_n)/K_p$ – an imperceptible change for $\overline{DP}_n \gg 1$. However, the weight average degree of polymerization increases by a factor of nearly 2.

---

* The simplified assumption of $K_1 = K_2 = \ldots = K_i = K_p$ is made again. However, this does not affect the essence of the argument to be presented

The kinetics of such a distribution change were discussed by Brown and Szwarc[195] and rigorously treated by Miyake and Stockmayer[196]. The approximate treatment of Brown and Szwarc gives

$$d(\overline{DP}_n)/dt = 2k_d/\overline{DP}_n(1 - x_0\overline{DP}_n)$$

where x_0 is the degree of polymerization of that living polymer which may grow but not degrade, e.g. for polymers of the kind $IM_{i-1}M^*$ $x_0 = 1$. The conversion is very slow for a high degree of polymerization, especially whenever k_d is low.

II.5. Conversion of Cyclic Monomers into Linear Polymers

Two conceptual reactions could be visualized for every cyclic monomer: a conversion into a cyclic oligomer:

n (XY) ⟶ ⌐X–YX–YX .... –Y⌐

or a conversion into a linear, high-molecular weight polymer:

n (XY) ⟶ 'X–YX–YX....–Y'

Moreover, any cyclic oligomer conceptually could be converted into a linear, high-molecular weight polymer like the one formed from a small cyclic monomer.

Let ΔG_i denote the free energy change of conversion of one base-mole of a cyclic i-mer into one base-mole of a linear, high-molecular weight polymer*. For $\Delta G_i < 0$, a conversion of the oligomer into a linear polymer could proceed spontaneously, provided that a mechanism for the conversion is available. However, for $\Delta G_i > 0$, thermodynamic restriction prevents the polymerization; but then a spontaneous degradation of a linear polymer into cyclic i-mers is allowed and it takes place whenever a mechanism for such a process is available. In fact, interconversion of one cyclic oligomer into another, e.g. $4(SiMe_2O)_3 \rightleftharpoons 3(SiMe_2O)_4$, is expected to proceed through the intermediacy of a linear polymer. Since living polymers are capable of growing as well as of degrading, this kind of conversion could take place in such systems. For example, ethylene oxide polymerizes and yields a high-molecular weight polyglycol. The latter could be degraded into dioxane – a dimer of the oxirane. Conversion of oxetane into a linear living polymer is readily achieved; but the latter degrades, yielding the respective cyclic tetramers[201]. A spectacular phenomenon was described by Goethals[199]. Polymerization of thiiran into a living linear polymer readily takes place; a three-membered cyclic sulphonium ion forms its growing end-group. However, a slow conversion of that end-group into an inert 12-membered cyclic sulphonium end-group transforms a living polymer into a dormant one, and eventually the dormant polymers degrade into 12-membered cyclic tetramers. Another remarkable result was reported by Brown and Slusarczuk[202]. Polymerization of

---

* Note ΔG_i is concentration dependent

the cyclic tetramer $(SiMe_2O)_4$ could be initiated by potassium hydroxide. Eventually, an equilibrated mixture of linear polymers and of a variety of cyclic oligomers is formed. The cyclic oligomers up to $n = 25$ were isolated and identified. Their concentrations in the equilibrated solution were determined, thus providing the first and most extensive quantitative test of the Jacobson-Stockmayer treatment of ring-chain equilibria[203]. Other examples of similar phenomena were reported in the literature[198].

In conclusion, a conversion of a cyclic monomer into a living linear polymer is thermodynamically allowed, provided that the reaction reduces the free energy of the system. However, the ultimate state of equilibrium corresponds to a mixture of cyclic oligomers and living linear polymers in appropriate proportion.

II.6. Equilibria in Living Poly-Trioxepane System

An interesting system related to those discussed previously was investigated by Schulz et al.[200], namely a cationic polymerization of trioxepane. This monomer may be treated as a cyclic co-polymer of ethylene oxide (E) and two molecules of formaldehyde (M)

Its cationic polymerization yields a living polymer terminated by $-O\dot{=}CH_2$ group, whereas the energetically improbable $-OCH_2CH_2^+$ group is never formed. Electrophilic attack of the $-O\dot{=}CH_2$ on oxygens of trioxepane yields two kinds of terminal sequences, either

$-OCH_2OCH_2CH_2OCH_2O\dot{=}CH_2$ (addition to oxygens 1 or 5)

or

$-OCH_2OCH_2OCH_2CH_2O\dot{=}CH_2$ (addition to oxygen 3).

The first sequence is briefly denoted as $-MEMM^+$; second, as $-MMEM^+$. Both are approximately equally probable because the statistical factor of 2 due to the presence of two identical oxygens (1 and 5) is balanced by the factor of 2 arising from the availability of two equivalent modes of opening acetal bonds following the attack on oxygen 3.

Since the two modes of addition are equally probable, the resulting polymer has a random structure with hetero-diads such as MEMEMM and EMMMEM as well as homodiads, viz. EMMEMM and MEMMEM. Its degradation may yield other monomers than trioxepane. The latter results from the reactions

$-MEMM^+ \rightarrow -M^+ +$ trioxepane

and

$-MMEM^+ \rightarrow -M^+ +$ trioxepane.

However, $-MEMM^+$ may also yield $-MEM^+$ and formaldehyde, while $-MMEM^+$ may yield $-MM^+$ and dioxolane (cyclic EM = $\underline{O}CH_2CH_2O\underline{C}H_2$). Both formaldehyde and dioxolane act as monomers in their own right and could be added to an $-M^+$ group. Thus four kinds of terminal sequences could be formed: (a) $-MEMM^+$, (b) $-MMEM^+$, (c) $-MEM^+$, and (d) $-MM^+$. Of course, class (d) includes class (a), while class (c) includes class (b).

Schulz' studies of cationic trioxepane polymerization demonstrated that the system rapidly approaches a state in which the living polymers cease to grow while the concentrations of the three monomers, trioxepane, dioxolane, and formaldehyde approach some stationary values as illustrated by Fig. 12.

The following treatment accounts for the behavior of this system[204]. Its final state is not that of equilibrium but rather of a stationary state. We could visualize a genuine equilibrium between each monomer and its living, perfectly uniform, high-molecular weight polymer. For example, the concentration of trioxepane in equilibrium with its living, uniform homo-polymer, i.e.

$$... EMM \cdot EMM \cdot EMM^+ \rightleftharpoons ... EMM \cdot EMM^+ + \text{trioxepane}, \qquad K'_T$$

n-mer $\qquad\qquad\qquad\qquad$ $(n-1)$-mer

$$... MEM \cdot MEM \cdot MEM^+ \rightleftharpoons ... MEM \cdot MEM^+ + \text{trioxepane}, \qquad K''_T$$

n-mer $\qquad\qquad\qquad\qquad$ $(n-1)$-mer

is equal to K'_T or K''_T. At this stage we allow for a distinction between K'_T and K''_T, but it will be shown later that $K'_T = K''_T$. In an analogous way the equilibrium concentrations of dioxolane and of CH_2O in contact with their living, uniform, high-molecular weight polymers are given by the equilibrium constants K_D and K_F, respectively, i.e.

$$... EM \cdot EM \cdot EM^+ \rightleftharpoons ... EM \cdot EM^+ + \text{dioxolane}, \qquad K_D$$

n-mer $\qquad\qquad\qquad\qquad$ $(n-1)$-mer

or

$$... MMM^+ \rightleftharpoons ... MM^+ + CH_2O, \qquad\qquad K_F$$

n-mer $\qquad$ $(n-1)$-mer .

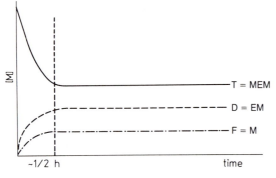

Fig. 12. The concentrations of trioxepane, MEM, oxolane, ME, and formaldehyde M, as functions of time. The variations are due to polymerization of trioxepane and the degradation of the resulting polymers into dioxolane and formaldehyde. Note the simultaneous approach to the final stationary state of the system

These equilibrium constants, and therefore the monomer equilibrium concentrations, are independent of the mechanism of polymerization and of *the initial concentration of the monomer,* being uniquely determined by the standard free energy of the monomer and of the monomeric segment of the respective *uniform,* high-molecular weight polymer.

The situation is more complex in the system under discussion. In conventional systems every living macromolecule can add its monomer as well as depropagate into its monomer. In the present system every living polymer may add any of the three monomers, trioxepane, dioxolane, or CH_2O; however, only a fraction of living macromolecules formed in the process may depropagate into a specified monomer. For example, polymers terminated by a sequence ...MEMM$^+$ may dissociate into trioxepane or CH_2O but not into dioxolane; those terminated by ...MMEM$^+$ could yield on dissociation trioxepane or dioxolane but not CH_2O. Polymers terminated by ...MMM$^+$ dissociate only into CH_2O, while those terminated by ...EMEM$^+$ yield on dissociation only dioxolane.

In what follows it is assumed that the rates of propagation depend only on the nature of the added monomer, while the rates of depropagation in which monomer X is expelled and a new ...M$^+$ ending is left, viz.,

$$...MX^+ \rightleftharpoons ...M^+ + X$$

depend only on X and not on the nature of units preceding the eventually formed, terminal ...M$^+$. For example, ...EM$^+$ and ...MM$^+$ are supposed to be equally reactive in any addition, and ...EMM$^+$ and ...MMM$^+$ are assumed to dissociate with the same rate into CH_2O. Let us denote by $f_{MEMM^+}, f_{MMEM^+}, f_{MEM^+}, f_{MM^+}$ the mole fractions of polymers terminated by ...MEMM$^+$, ...MMEM$^+$, ...MEM$^+$, and MM$^+$, respectively. These mole fractions become constant when the system attains its stationary state. Of course, $f_{MEM^+} + f_{MM^+} = 1$, while $f_{MEMM^+} < f_{MM^+}$ and $f_{MMEM^+} < f_{MEM^+}$.

At a constant concentration of living polymers, the pseudo-first order rate constants for each monomer addition remain the same whether the polymers are or are not uniform. However, their rate of dissociation into a specified monomer is smaller in the system discussed here than in that involving the appropriate uniform polymers. While each uniform polymer may dissociate into its monomer, only a fraction of non-uniform polymers, i.e. those possessing the appropriate ending, is capable of degrading into that monomer. Therefore, the stationary concentrations of the respective monomers are given by their "true" equilibrium concentrations multiplied by the mole fractions of those with the required ending, viz.

$$[trioxepane]_{st} \quad = K'_T f_{MEMM^+} = K''_T f_{MMEM^+}$$

$$[dioxolane]_{st} \quad = K_D f_{MEM^+}$$

$$[CH_2O]_{st} \quad\quad = K_F f_{MM^+} \; .$$

Let us idealize the system and assume that all the macromolecules are of high molecular weight although their molecular weight distribution may be arbitrary. Moreover, we will assume that their compositions are identical. By composition we understand the ratio r of E units to M units in the polymers. Enthalpy of a polymer terminated by M$^+$ and composed of n E units and m M units (excluding the terminal M$^+$) is not affected by any permutation of the segments, provided that no two E's are adjoined. This is an obvious and necessary condition which is automatically fulfilled if the permutations involve the conceptual EM units and the residual M units. Hence, the number of *allowed* permutations of E's and M's leaving E's disjoined is given by the total number of permutations of n EM's and $(m - n)$ M's (M preceding $-M^+$) and the total number of permutations of $(n - 1)$ EM's and $(m - n + 1)$ M's (E preceding $-M^+$), viz. $(m + 1)!/n!(m - n + 1)!$. In a similar way, one finds the number of allowed permutations in polymers terminated by $-MM^+$ to be $m!/n!(m - n)!$, in polymers terminated by $-MEM^+$ to be $m!/(n - 1)!(m - n + 1)!$, and in polymers terminated by $-MEMM^+$ or $-MMEM^+$ to be $(m - 1)!/(n - 1)!(m - n)!$. Since each permutation is equally probable, one finds the mole fraction of polymers terminated by $-MM^+$, $-MEM^+$, $-MEMM^+$, and $-MMEM^+$ to be*

$$f_{MM^+} \quad = (m - n + 1)/(m + 1) \approx 1 - n/m = 1 - r,$$

$$f_{MEM^+} \quad = n/(m + 1) \approx n/m = r,$$

$$f_{MEMM^+} = f_{MMEM^+} = n(m - n + 1)/m(m + 1) \approx (n/m)(1 - n/m) = r(1 - r) .$$

Since $f_{MEMM^+} = f_{MMEM^+}$ and $K'_T = K''_T = K_T$, the pseudo-equilibrium concentrations of the monomers in contact with the living polymers having composition r is given therefore by:

$$[\text{trioxepane}]_{st} \quad = K_T r(1 - r)$$

$$[\text{dioxolane}]_{st} \quad = K_D r$$

$$[CH_2O]_{st} \quad \quad = K_F(1 - r) .$$

Note that the pseudo-equilibrium concentrations of the monomers depend on the composition of the polymer and therefore on the composition and concentration of the initial feed. For example, let us consider polymerization initiated in a solution of trioxepane having initial concentration C_0. The stoichiometric balance requires

$$\{C_0 - K_T r (1 - r) - K_D r\}/\{2 C_0 - 2 K_T r (1 - r) - K_D r - K_F (1 - r)\} = r,$$

i.e. r is determined by C_0. Generally, r is determined by the initial concentration of E units, C_E, and M units, γC_E, in the initial feed. Of course, $\gamma > 1$.

The detailed treatment of trioxepane polymerization serves to illustrate the complexities that could be encountered in living polymer systems and to outline the approaches helpful in unraveling the resulting problems.

---

* Note: n and m are very large numbers

III. Initiation of Anionic Polymerization

Anionic polymerization is initiated by processes that converting a monomer into the smallest species possessing the characteristic negatively charged end-group regenerated in each step of propagation. This results from transfer of negative charge to a monomer with simultaneous addition of some moiety derived either from an initiator or from another monomer.

Most of the early mechanistic investigations of anionic polymerization were concerned with reactions taking place in liquid ammonia. The system liquid ammonia-alkali metals will be dicussed first, followed by a review of heterogeneous reactions taking place on alkali or alkali-earth surfaces. Thereafter homogeneous electron-transfer processes and the addition of negative ions to monomer will be discussed. Finally some esoteric reactions, such as initiation by Lewis bases, charge-transfer complex initiation, etc. will be briefly reviewed.

III.1. Alkali Metal-Solvent Systems

The solubility of alkali metals in most of the solvents used in anionic polymerization is exceedingly low. Liquid ammonia and hexamethylphosphoric-triamide are exceptional in this respect; relatively high concentrations of the metals can be attained in these media. Several distinct species co-exist in alkali metal solutions. A minute but constant concentration of unionized atoms, Met, is maintained when a solvent is kept in contact with solid alkali metal. These in turn are in equilibrium with the products of their ionization:

$$\text{Met} \rightleftharpoons \text{Met}^+ + \text{solvated electron, e}^-, \qquad K_1, \qquad (1,1)$$

Further equilibrium is established between Met and e^-, i.e.,

$$\text{Met} + e^- \rightleftharpoons \text{Met}^-, \qquad K_2, \qquad (1,2)$$

yielding the negative alkali ions[63]. Finally, the ions are in equilibrium with ion-pairs,

$$\text{Met}^+ + e^- \rightleftharpoons \text{Met}^+, e^-, \qquad K_3, \qquad (1,3)$$

and

$$\text{Met}^+ + \text{Met}^- \rightleftharpoons \text{Met}^+, \text{Met}^-, \qquad K_4. \qquad (1,4)$$

Higher aggregates need not be considered since most of the investigated solutions are rather dilute.

Due to the common ion effect, the concentration of solvated electrons is depressed by the presence of ionized salts of the appropriate alkali. This in turn depresses the concentration of the negative alkali ions. However, as long as contact with the solid metal is maintained, the concentrations of the unionized atoms as well as of the ion-pairs remain unaffected. The latter are given by:

$$[Met^+, e^-]_e = K_1K_3[Met]_e$$

and

$$[Met^+, Met^-]_e = K_1K_2K_4[Met]_e^2 ,$$

the K's denoting the relevant equilibrium constants. At higher concentration of ions the above equations refer to the corresponding activities.

The equilibrium concentration of the unionized atoms depends mainly on the heat of sublimation of the pertinent metal; their solvation energy is probably small. Hence $[Cs]_e > [Rb]_e > [K]_e > [Na]_e$ and the equilibrium concentration of Li atoms is probably the smallest, since no Li hyperfine ESR lines were observed in solutions of metallic lithium.

The value of K_1 increases with increasing dielectric constant of the medium, but the ability of solvent molecules to be coordinated with cations is even more important in enhancing the degree of ionization. For example, K_1 is substantially greater in dimethoxy-ethane than in tetrahydrofuran, although the dielectric constants of both ethers are virtually identical. The bidentate nature of $CH_3OC_2H_4OCH_3$ molecules makes dimethoxyethane a much better coordinating agent than tetrahydrofuran.

The available data suggest the following order of solvent's capacity to ionize alkali metals, namely

liquid ammonia > hexamethylphosphoric-triamide > methylamine ~ ethylene diamine ~ dimethoxyethane > ethylamine > tetrahydrofuran ≫ diethylether.

The reverse order favors the formation of ion-pairs.

While solvated electrons and their ion-pairs seem to be the most abundant species in liquid ammonia and hexamethylphosphoric-triamide, the negative alkali ions and their ion-pairs predominate in ethereal solvents. In fact, pulse radiolysis of tetrahydrofuran solutions of alkali tetraphenylborides revealed that the e^-, Met^+ pairs, observed immediately after a pulse, rapidly decay, being converted into the corresponding Met^- ions[64].

Powerful cation solvating agents were discovered in recent years: the crown ethers and the kryptates. In their presence, the amount of dissolved alkali metals substantially increases[65]. Their association with cations,

$$Met^+ + \text{crown or kryptate} \rightleftharpoons Met^+, \text{crown (or kryptate)} , \tag{1,5}$$

usually yields 1:1 complexes, although 1:2 complexes have been observed in some systems, e.g., Cs^+, 2 crown or Ba^{2+}, 2 crown[66]. The increase of total concentration of cations requires an equivalent increase of concentration of anions in order to maintain the electric neutrality of the medium. Denoting by [X] the concentration of a crown or kryptate, we find

$$[e^-] + [Met^-] = \{K_1[Met]_e(1 + K_2[Met]_e)(1 + K_5[X])\}^{1/2} .$$

Since K_5 is large, the concentration of anions is proportional to the square root of the concentration of the complexing agent, provided that its concentration is not too low.

Further amplification of crown or kryptate action arises from their coordination with ion-pairs:

$$Met^+, e^- + X \rightleftharpoons (Met^+, X), e^- \tag{1,6}$$

and

$$Met^+, Met^- + X \rightleftharpoons (Met^+, X), Met^- . \tag{1,7}$$

Hence, $[Met^+, e^-] + [(Met^+, X), e^-] = K_1K_3[Met]_e(1 + K_6[X])$

and $[Met^+, Met^-] + [(Met^+, X), Met^-] = K_1K_2K_4[Met]_e^2(1 + K_7[X])$.

In fact, a crystalline salt Na^+, Cryptate, Na^- was prepared and its structure was confirmed by X-ray analysis[67].

III.2. Structure and Properties of Solvated Electrons, their Ion-Pairs, and of the Negative Alkali Ions

An electron trapped in a polar liquid is stabilized by creating a cluster of properly oriented solvent molecules, their dipoles generating a nearly spherically symmetric field within which the electron moves. Under such conditions the electron's energy is quantized. Therefore solvated electrons have characteristic absorption spectra consisting of an intense, very broad and structureless band extending from the visible to the near-infrared region and showing marked asymmetry on its high energy side. Presumably some variation of the clusters' dimensions is responsible for the broadness of these bands. The maximum of the absorption depends on the nature of solvent but not of alkali metal. For example, λ_{max} of solvated electrons appears at 630 nm in water, in liquid ammonia at 1480 nm, and in tetrahydrofuran at $\sim$2020 nm.

Oriented clusters of solvent molecules surrounding electrons have cavities in their middle, their size being interpreted as the dimension of solvated electrons. Such a cavitation is revealed by dilation of the medium upon injection of electrons as measured by ΔV[68].

$$\Delta V = V_{solution} - (V_{solvent} + V_{metal}) \text{ per gram atom of metal.}$$

From the value of ΔV the radius of cavity can be calculated, e.g., 4.5 A for electrons in ammonia. The cavitation is responsible for the exceedingly low density of lithium solutions, e.g., saturated lithium solution in liquid ammonia is the lightest liquid existing at ambient temperature, its density being 0.48 g/ml.

Solvated electrons are mobile and conduct in an electric field. Their mobility is higher than that of other ions in the same medium although not by a large factor. This implies that electron hopping mechanism, as well as migration of the entire cavity, contribute to their conductance. In this respect, solvated electrons should be distinguished from the fast moving "dry" electrons formed in the initial stage of ionization occurring during radiolysis[69]. Solvated electrons are paramagnetic; their ESR spectrum reveals a single, very sharp peak at the expected g value of a free spin.

The spectra of e^-, Met^+ resemble those of solvated e^-. Their absorption maxima are shifted to shorter wavelength; the shift *depends* on the nature of cation[70]. For the sake of illustration, the spectra of $alkali^+$-e^- pairs in tetrahydrofuran are shown, together with that of e^-, in Fig. 13. The similarity of those spectra argues for the ion-pair structure of these species rather than the centrosymmetric expanded atom model proposed by early workers. The equilibrium

$$e^-, Met^+ \rightleftharpoons Met$$

resulting in a loss of solvation energy of the cation favors the ionized species. Nevertheless, the magnetic properties of these solutions reveal a rapid and reversible interconversion between the pair and the atom.

Like solvated electrons, the pairs and the atoms are paramagnetic. Their ESR spectra reveal hyperfine structure arising from the interaction of the electron's spin with the spin of alkali nucleus. The coupling constant increases enormously with rising temperature, an increase unmatched by those observed in any other systems. This observation is consistent with the proposed dynamic equilibrium between the pairs and the atoms, higher temperatures favor the unionized atom. Apparently, the rate of conversion is high

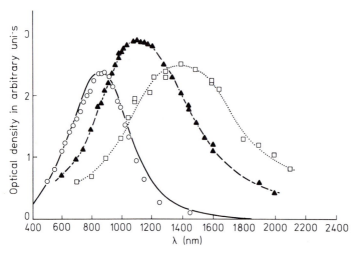

Fig. 13. Spectra of e^-, Cat^+ in THF. Observed immediately after pulsing a THF solution of the appropriate tetraphenyl boride salt. Technique – pulse radiolysis. From left to right: e^-, Na^+, e^-, K^+, e^-, Cs^+

on the ESR time scale, resulting in an average spectrum. However, in certain cases, two sets of lines corresponding to two different coupling constants are observed, i.e. the rate of interconversion is substantially slower in those systems than in the others.

The negative alkali ions are diamagnetic. They were incorrectly described in older papers as diamagnetic $(e^-)_2$. Interestingly, the existence of negative alkali ions in the gas phase was postulated by Russian physicists in the early 1930's, but the prejudice of chemists prevented the acceptance of their existence until recent years. The ion-pairs Met$^-$, Met$^+$ are stoichiometrically identical with diatomic molecules Met$_2$. In the gas phase, the diatomic Met$_2$ molecules as well as Met$_2^+$ ions are observed; however, their existence in solutions is unlikely.

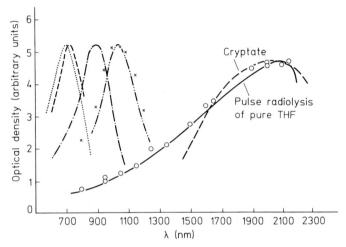

Fig. 14. Spectra of negative ions of alkali metals in THF. Observed 100 nsec. after pulsing a THF solution of the appropriate tetraphenyl boride salts. From left to right: Na$^-$, K$^-$, Cs$^-$. *Solid line:* spectrum of e$^-$ in THF. Obtained by pulsing pure THF. *Dashed line:* spectrum of e$^-$ in THF in the presence of 2,2,2-kryptate

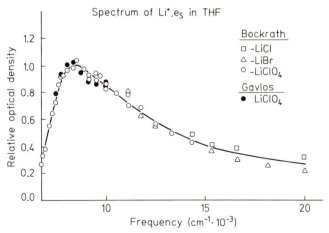

Fig. 15. Spectrum of e$^-$, Li$^+$ in THF. Recorded after pulsing THF solution of LiCl □, LiBr △, and LiClO$_4$ ○

The absorption spectra of negative alkali ions are sharper than those of the corresponding ion-pairs and their maxima are slightly shifted towards shorter wavelengths[71]. For the sake of illustration, the spectra of Na^-, K^-, and Cs^- in tetrahydrofuran are shown in Fig. 14. The spectrum of Li^- was not observed, although the spectrum of Li^+, e^-, shown in Fig. 15, was reported by Dorfman[72].

III.3. Reactivity of Alkali Metal Solutions

The extremely high reactivity of alkali metal solutions calls for the utmost care in their handling. They must not be exposed to the air; indeed, solutions such as potassium in hexamethylphosphoric-triamide are pyrophoric. Moisture and other adsorbed impurities have to be meticulously removed from all surfaces that come in contact with these reagents. A relatively simple technique to achieve this goal is described elsewhere[141].

The stability of alkali metal solutions depends on the nature of the solvent and its degree of purity. Solutions in hexamethylphosphoric-triamide are relatively stable, although some decomposition leading to the formation of alkali hydrides occurs after a few days. Liquid ammonia solutions, when kept at low temperature, evolve hardly any hydrogen. Apparently, the direct reaction, $NH_3 + e^- \rightarrow NH_2^- + H$, does not take place. However, solvated electrons react rapidly with ammonium ions formed through the self-ionization of the solvent, i.e., $NH_4^+ + e^- \rightarrow NH_3 + H$[73]. The slow rate of hydrogen evolution is governed, therefore, by the rate of self-ionization; hence, it is independent of the concentration of alkali metal and is retarded by amide ions. A similar sequence of reactions accounts for the formation of hydrogen arising from addition of water or alcohol to alkali metal solutions in liquid ammonia. The equilibrium

$$ROH + NH_3 \rightarrow RO^- + NH_4^+$$

substantially increases the concentration of ammonium ions causing enhanced evolution of gas. However, the rate of H_2 evolution is not influenced by the concentration of alkali metal, and the reaction is retarded by alkoxide ions formed in its course. Both observations confirm the proposed mechanism.

It is interesting to realize that the above mechanism predicts the formation of solvated electrons in solutions of amides in liquid ammonia or hydroxides in water when these are pressurized by hydrogen gas. The equilibrium,

$$NH_2^- \; (or \; OH^-) \; + \frac{1}{2} \, H_2 \; \rightleftharpoons \; NH_3 \; (or \; H_2O) + solvated \; e^- \; ,$$

has to be established, whatever its mechanism. Indeed, solvated electrons were detected under these conditions by the ESR technique[74].

Electron-metal$^+$ ion-pairs are more reactive than the free solvated electrons in reactions yielding alkoxide or amides. Apparently, the charge transfer required for the formation of those anions is facilitated by the proximity of Met^+ cation which becomes coordinated with the oncomming X–OR reagent. Since, in contrast to liquid ammonia, ion-pairs are the abundant species in organic amine or ether solutions, these directly react with water or alcohol. The reaction has a second order behavior and its rate is not retarded by hydroxide ions.

Alkoxide formation is responsible for the slow destruction of alkali-metal solutions in ethers. For example,

$$e^-, Met^+ + CH_3OCH_2CH_2OCH_3 \rightarrow CH_3O^-, Met^+ + \cdot CH_2CH_2OCH_3 .$$

However, stability of such solutions is greatly improved by the addition of crown ethers or kryptates. Their coordination with the tight ion-pairs results in formation of the less reactive loose pairs. A similar phenomenon is observed in the protonation of radical-anions dissolved in ethers[106].

Solvated electrons, their ion-pairs and the negative alkali ions are powerful reducing agents capable of initiating anionic polymerization. Homogeneous initiation is practical when their concentration is sufficiently high, say 10^{-5} M or more. This condition is readily fulfilled in the presence of powerful solvating agents like the crowns or kryptates.

Only a few examples of polymerization initiated by alkali-metal solutions were reported in the literature. Overberger et al.[75] investigated polymerization of methacrylonitrile initiated in liquid ammonia solutions of lithium at $-78\,^\circ$C. The conversion was quantitative and completed in a few seconds. The molecular weight of the polymer was given by the ratio 2[monomer]/[dissolved Li], an observation implying the formation of living polymers by electron-transfer. Presumably the association of Li$^+$ with the respective living end-group retards the termination caused by the proton-transfer from the solvent, although the propagation is still allowed. In contrast to the solution of lithium metal, the solution of lithium amide was ineffective in that solvent, most probably because the amide salt consists of unreactive, hardly dissociated, tight ion-pairs.

Polymerization of methacrylonitrile by liquid ammonia solutions of potassium led to a polymer having constant molecular weight independent of the [monomer]/[dissolved K] ratio[76]. Probably the association of the living end-groups with this cation is less pronounced. Consequently, the proportion of free growing anions, present in equilibrium with ion-pairs, is high and these are protonated by ammonia yielding potassium amide which acts as initiator. Indeed, potassium amide in liquid ammonia, in contradistinction to lithium amide, initiates polymerization of methacrylonitrile.

Blue solutions of potassium in dimethoxyethane or tetrahydrofuran initiate polymerization of styrene[77]. With the former solvent the reaction was quantitative, both at 0° and $-70\,^\circ$C, yielding living polystyrene. Termination was observed in tetrahydrofuran; presumably it was caused by some slowly reacting impurities. The authors attributed this initiation to diamagnetic $(e^-)_2$; however, the diamagnetic blue species are the K$^-$ anions which act as the electron-transfer agents.

An attempt to polymerize methyl-vinyl sulphone in diglyme solution of potassium was unsuccessful[78]. Only dimers, instead of polymers, were formed. Apparently K$^-$ anions react with the monomer according to the equation

$$2\,CH_2:CH \cdot SO_2CH_3 + K^-, K^+ \rightarrow 2\,\overline{CH_2:CH \cdot SO_2} \cdot CH_3, K^+ .$$

The resulting radical anions are protonated by the unreacted sulphone in a process yielding radicals

K^+, $CH_2^- : CH \cdot SO_2 \cdot CH_3 + CH_2 : CH \cdot SO_2 \cdot CH_3 \rightarrow$
$\cdot CH_2 \cdot CH_2 \cdot SO_2 \cdot CH_3 + CH_2 : CH \cdot SO_2CH_2^-$, K^+ ,

and their dimerization accounts for the ultimately formed product.

Rapid and quantitative polymerization of 4-vinyl-pyridine in liquid ammonia solution of sodium at $-50\,°C$ was reported by Laurin and Parravano[79]. A high molecular weight and apparently living polymer was obtained in less than a minute.

III.4. Heterogeneous Reactions on Alkali and Alkaline-Earth Surfaces

Since rates of heterogenous reactions are proportional to the surface area of solid reagents, it is advantageous to maximize the area as far as possible. In laboratory reactions involving sodium, potassium, rubidium or cesium, the metal is deposited by evaporating it from a side container onto the surface of a reactor to form a mirror. Pyrex tubes can be used as containers since the evaporation is sufficiently fast at temperatures not exceeding the softening point of pyrex glass. However, for deposition of calcium, strontium or barium one has to utilize quartz tubes linked to a pyrex reactor through graded-joints. Gas-torch heating is sufficient in deposition of alkali metal, but it is impractical in the sublimation of alkaline-earth metals which require electric heating. Mirrors of lithium cannot be deposited by this procedure since lithium reacts with glass at elevated temperatures causing it to crack. Nevertheless, lithium mirrors could be deposited on the walls of a reactor by introducing into it a solution of lithium in liquid ammonia and evaporating the solvent. Such mirrors are often contaminated by lithium amide.

It is important to keep the metal surface clean and free of layers of oxide or hydroxide. Shiny mirror surfaces obtained by vacuum deposition need no further purification, whereas chunks of lithium or other alkali metals have to be freed of oxide layers. This is achieved by keeping them in evacuated containers under tetrahydrofuran solution of aromatic hydrocarbons, e.g., anthracene. During the reduction of the aromatic hydrocarbon to its radical-ion or dianion the oxide peels off exposing a shiny surface of the clean metal. The resulting solution with the oxide dispersed in it is decanted to a side bulb, the solvent distilled back by cooling the reactor and then, after shaking, decanted again. This operation is repeated a few times, and finally the side bulb is chilled with liquid nitrogen and then sealed off from the reactor.

For large scale operation the metal is melted under high boiling inert hydrocarbon, vigorously stirred or shaken, and then suddenly cooled to produce fine granules of alkali metal. Unfortunately, this method is not feasible when dealing with metallic lithium since its melting point exceeds $200\,°C$. Although commercial emulsions of lithium are available, their use in precise experiments is not recommended because they contain emulsifiers. Liquid sodium-potassium alloy may be conveniently used in preparatory work. Stirring increases the surface area and keeps it renewable; the resulting products are only slightly contaminated by sodium since potassium is much the more reactive metal.

An elaborate technique yielding pyrophoric dispersions of alkaline-earth metals was described by Francois et al.[80]. Vapor of a continuously boiling inert hydrocarbon is circulated through a tube into which the vapor of alkaline-earth metal is injected via a nozzle. The metal, condensed by the diluent, forms a fine dispersion collected in the boiler, whereas the hydrocarbon is recycled through the apparatus.

In heterogeneous reactions one deals with two kinds of systems. A solution of an aromatic hydrocarbon, (A), in contact with alkali or alkaline-earth metal reaches a state of equilibrium:

Solid metal (Met) + solution of A $\rightleftharpoons$ solution of A$^{\cdot-}$, Met$^+$ (or A^{2-}, 2 Met$^+$)[81] .

The ultimate ratio [A$^{\cdot-}$, Met$^+$]/[A] or [A^{2-}, 2 Met$^+$]/[A] is independent of the initial concentration of A. For most systems it increases with decreasing temperature since usually the reaction is exothermic. The value of that ratio depends on the activity of the solid metal, the solvation power of the medium, and the reduction potential of A. The activity of the metal is determined by its heat of sublimation and the ionization potential of the resulting atom. The solvation power of the medium determines the free energy of solvation of the pertinent cation as well as of the resulting A$^{\cdot-}$ anion. The I_p of metal atom in conjunction with electron affinity of A determines the reduction potential of A. In low dielectric media, which are of the greatest interest for us, the ratio [free A$^{\cdot-}$]/[A$^{\cdot-}$, Met$^+$] is very low, i.e., virtually all the anions are in the form of ion-pairs. The equilibrium constant of ion-pairing comes, therefore, as another term contributing to the calculation of the [A$^{\cdot-}$, Met$^+$]/[A] ratio. The above treatment applies to those aromatic hydrocarbons that neither dimerize nor undergo any further reactions. These therefore form a useful class of homogeneous initiators of anionic polymerizations induced by electron transfer, provided that their electron affinity is not too high.

As implied by the above treatment, the conversion of A into A$^{\cdot-}$, Met$^+$ is never quantitative. For example, reduction of naphthalene in tetrahydrofuran by sodium yields about 95% of its radical anions at ambient temperature, whereas the yield in diethyl ether is very low, less than 1%. The sodium reduction of biphenyl, a hydrocarbon of lower electron affinity than naphthalene, results only in about 20% conversion in tetrahydrofuran, although nearly 100% reduction may be accomplished in dimethoxyethane. Since these processes lead to equilibria, it is immaterial whether they take place on the surface of the metal, being followed by desorption of the product, or whether they proceed homogeneously in a solution saturated by metal atoms and their ionization products, these being replenished from the bulk of the solid metal as the reduction proceeds.

The relation is different when the initial reaction is followed by a virtually irreversible process. For example, reduction of 1,1-diphenyl-ethylene yields radical-anions which subsequently dimerize. The dimerization is virtually irreversible; its rate constant was recently determined by the flash-photolysis technique[82] and shown to vary from 1×10^8 for the Li$^+$ salt to 30×10^8 M^{-1} s^{-1} for the Cs$^+$ salt. The irreversibility of dimerization makes the conversion quantitative in spite of the relatively low electron affinity of the ethylene derivative.

An interesting process is observed in the reduction of α-methylstyrene by sodium-potassium alloy[83]. Dimerization of the initially formed radical-ions yields dianions. Their subsequent growth leads to a relatively high-molecular weight, living poly-α-methylstyrene. However, since contact with the metal is retained, the residual monomer – co-existing in equilibrium with the living polymers – continues to be reduced, and thus additional dimers are formed. Depletion of the monomer results in degradation of living polymers since the equilibrium concentration of the monomer has to be maintained. Consequently, the high-molecular weight polymers formed in the early stages of the reaction eventually disappear, and ultimately all the monomer initially introduced into the system is converted into dimeric dianions,

$$K^+, \overline{C}(Ph)(CH_3) \cdot CH_2 \cdot CH_2 \cdot \overline{C}(Ph)(CH_3), K^+ .$$

The latter are useful as homogeneous initiators of anionic polymerization.

A similar sequence of reactions takes place when sodium metal replaces the alloy, but then tetrameric dianions, instead of dimeric ones, are the main products of that process. Apparently, the slowness of reduction by sodium metal is responsible for this result. The tetramer was described in earlier papers as a tail-to-tail, head-to-head, tail-to-tail bonded oligomer[84]. Subsequent NMR studies[85] have conclusively shown that they result from the addition of two monomeric molecules to the dimeric dianions, i.e.

$$Na^+, \overline{C}(Ph)(CH_3) \cdot CH_2 \cdot C(Ph)(CH_3) \cdot CH_2 \cdot CH_2 \cdot C(Ph)(CH_3) \cdot CH_2 \cdot \overline{C}(Ph)(CH_3), Na^+ .$$

The tetramers are also useful initiators of homogeneous, anionic polymerization.

The drawback of initiation by solid alkali metals is revealed by the above examples. Undoubtedly, the initiation occurs on the surface of the metal and yields adsorbed living oligomers. The latter are eventually desorbed and continue their growth in solution. Since the propagation is fast while the slow, heterogeneous initiation continues, the molecular-weight distribution of the ultimately formed polymer is rather broad. Its average molecular weight is unpredictable because its value depends on the surface area of the metal – a factor not readily reproduced.

Evidence of monomer adsorption on a metal surface is provided by the extensive studies of Richards and his co-workers[86]. As is well known, alkyl bromides in tetrahydrofuran vigorously react with alkali metals, say lithium, yielding the Wurtz coupling products. The violent reaction slows down on addition of aromatic monomers like styrene, and the nature of the products is drastically changed[87]. For example, when an equimolar mixture of ethylbromide and styrene reacts in tetrahydrofuran with metallic lithium, the alkyl capped dimer, $C_2H_5 \cdot CH(Ph) \cdot CH_2 \cdot CH_2 \cdot CH(Ph) \cdot C_2H_5$ forms 90% of the products, about 5% appear as $C_2H_5 \cdot CH_2 \cdot CH(Ph) \cdot C_2H_5$, the remainder being a mixture of an alkyl capped trimer and butane. The tail-to-tail structure of the capped dimer was demonstrated by the NMR technique.

It seems that styrene and ethylbromide compete for the sites on the lithium surface. The adsorption of styrene possessing easily polarized π electrons is favored over that of ethylbromide. Hence, Wurtz coupling is hindered, while the electron transfer to the adsorbed styrene yields its radical-anions, and their mobility on the surface allows for their dimerization. The dimeric dianions are eventually desorbed, and since their reaction with ethylbromide is faster than propagation, the ethyl capped dimers are the main products.

Further support for the proposed mechanism is provided by the results of experiments involving phenylbromide instead of ethylbromide[88]. The polarizable π electrons of this aryl compound allow it to effectively compete with styrene for the sites on the lithium surface and thus the Wurtz coupling reaction becomes dominant. Similar results were obtained with ethyltosylate. Although the reaction of tosylate with living polystyrene is rapid and quantitative, yielding ethyl capped polymers, its reaction with the monomer and metallic lithium produces only 10% of the ethyl capped polymers, the remainder being evolved as butane. Again, the aromatic nature of tosylate allows it to compete with styrene for the lithium sites.

An interesting extension of this picture is provided by the behavior of p-xylylene dibromide[89]. With butadiene as monomer and tetrahydrofuran as solvent, their reaction on metallic lithium leads to an unusual co-polymer,

$$-(BD)_n-CH_2 \cdot C_6H_4 \cdot CH_2 \cdot CH_2 \cdot C_6H_4 \cdot CH_2-(BD)_m- \quad (BD\text{-butadiene moiety}) .$$

Its composition is determined by the initial ratio of the reagents in the feed. NMR analysis confirmed the above structure and showed that the p-xylylene moieties in the polymeric chains exclusively exist as dimers.

Unexpectedly, the vicinyl dihalides react differently[90]. For example, an equimolar mixture of styrene and 1,2-dibromoethane reacts on lithium metal yielding a head-to-head, tail-to-tail polystyrene, ethylene and lithium bromide. Apparently, the adsorbed styrene is reduced and dimerized to the dianions,

$$^-CH(Ph) \cdot CH_2 \cdot CH_2 \cdot CH(Ph)^- ,$$

and the latter rapidly react with the dibromide yielding the unconventional polymer with elimination of ethylene

$$n \{Li^+, {}^-CH(Ph) \cdot CH_2 \cdot CH_2 \cdot CH(Ph)^-, Li^+ + Li^+, {}^-CH(Ph) \cdot CH_2 \cdot CH_2 \cdot CH(Ph)^-, Li^+\}$$
$$\downarrow CH_2BrCH_2Br$$
$$-[CH(Ph) \cdot CH_2 \cdot CH_2 \cdot CH(Ph)-CH(Ph) \cdot CH_2 \cdot CH_2 \cdot CH(Ph)]_n- + n C_2H_4 + 4n LiBr .$$

Other linking agents were investigated, e.g. dibromo-dimethyl-silane and dichlorophenylphosphine. Interesting products were obtained with diepoxides[91] namely

$$^-M \cdot M \cdot CH_2 \cdot \underset{\underset{OLi}{|}}{CH} \cdot R \cdot \underset{\underset{OLi}{|}}{CH} \cdot CH_2 \cdot M \cdot M \cdot {}^- \quad etc.,$$

yielding the respective poly-ols on hydrolysis.

As expected, different monomers are adsorbed to a different degree on alkali metal surfaces. Such a differentiation in adsorption capacity accounts for some confusing results reported in the literature. The homogeneous anionic co-polymerization of an equimolar mixture of styrene and methyl-methacrylate initiated by butyl lithium yields a virtually pure homo-poly-methyl-methacrylate containing about 1% styrene. However, when a lithium metal dispersion was used for the initiation, the resulting co-polymer contained a significant proportion of styrene. Tobolsky, who was first to report these results[92], argued that electron-transfer from metallic lithium yielded radical-anions of the monomers and they, in turn, initiated simultaneously a radical and anionic polymerization[93]. The product was assumed to be a block polymer, possessing a virtually styrene-free block of polymethyl-methacrylate arising from the anionic polymerization, and a block of an about 50:50 random co-polymer expected to be formed by a radical copolymerization. This suggestion was disproved by Overberger who repeated the experiments and demonstrated by the NMR technique that at low degrees of conversion the resulting polymer was composed of a short block of homo-polystyrene followed by a block of homo-polymethyl-methacrylate[94]. The preferential adsorption of styrene on lithium surface accounts for these findings. The initial surface polymerization involves the adsorbed styrene. The poly-styrene chains of higher degree of polymerization are desorbed – the gain in the conformational entropy provides the driving force. The desorbed living poly-styrene initiates then a homogeneous anionic polymerization of methyl-methacrylate, yielding a block of poly-methyl-methacrylate.

Another aspect of heterogeneous initiation on alkali metal surfaces deserves comments. Electron-transfer from a metal to an adsorbed monomer does not require simultaneous removal of a cation from the metal's lattice. The positively charged surface acts as a "counter-ion" causing strong attachment of the anions to the surface. Nevertheless, the adsorbed species may dimerize and propagate yielding oligomers still attached to the surface. The large gain in conformational entropy of those oligomers, arising from the freedom they acquire in solution, leads eventually to their desorption. In this process cations have to be extracted from the lattice of a metal, this being most difficult for lithium. The extraction is facilitated by solvation of the cations, and hence the higher the solvation power of the medium the easier the desorption which then can take place at lower degree of polymerization of the adsorbed oligomers. Indeed, as demonstrated by Overberger, the size of the polystyrene block increases with decreasing solvation power of the medium. For example, the reactions performed with lithium metal dispersion in the absence of ethers yields co-polymer with 28% of styrene at 1% conversion[94]. On the other hand, the facile removal of Na^+ cations from the sodium lattice might explain the formation of homo-poly-methyl-methacrylate in similar experiments involving sodium dispersion instead of lithium dispersion.

Removal of metal cation from lattices of alkaline-earth is very difficult. Consequently, initiation is extremely slow and the concentration of radical-anions released into the solvent can be very low. Under such conditions radical growth might compete with their dimerization. The reported formation of random co-polymers of styrene and methyl-methacrylate when the mixture of these monomers reacts with dispersions of alkaline-earth metals might be evidence for such a polymerization[95]. Confirmation of these observations is desirable.

Anionic polymerization of styrene, butadiene, isoprene and methyl methacrylate initiated by alkali metals inserted in graphite layers was investigated by Loria, Merle et al.[446]. The initiating species were described $AlkC_8$, $AlkC_{12}$, $AlkC_{24}$, $AlkC_{32}$, etc., where Alk stands for Li, Na or K. The initiation as well as the ensuing propagation depend on the rate of diffusion of the monomer into the graphite layers which swell as the reaction proceeds. The system is complex and the results are not readily interpreted. Catalytic action of graphite intercalated compounds was extensively reviewed by Ebbert[491].

III.5. Homogeneous Initiation by Radical-Anions

Radical-anions, $A^{\bar{}}$, act as useful electron donors converting monomers, M, into their radical-ions:

$$A^{\bar{}} + M \rightleftharpoons A + M^{\bar{}} \cdot .$$

Since the radical-anions derived from monomers undergo rapid and virtually irreversible dimerization, the electron-transfer is quantitative even for an unfavorable equilibrium. Addition of the monomer to the resulting dimeric-dianions is then the first step of anionic propagation.

This kind of initiation is rapid; the dimerization of $M^{\bar{}}$, rather than the electron transfer, is its rate determining step. A numerical example is illuminating. Consider the initial conditions of a 1 M THF solution of a monomer with 10^{-4} M · of an initiator $A^{\bar{}}, Cat^+$. For K_{tr} as low as 10^{-5}, the electron-transfer equilibrium, established within a few μsec*, results in 27% of $A^{\bar{}}, Cat^+$ being converted into monomer$^{\bar{}}, Cat^+$. Even for k_d as low as $10^6 \ M^{-1}s^{-1}$, it takes less than 0.2 s to produce 95% of all the dimeric dianions expected in the quantitative conversion. In that time less than 50 monomer molecules are added to each of the formed growing centers, provided the propagation constant k_p, is not larger than 250 $M^{-1}s^{-1}$ – a rather high value, while 10^4 molecules are added at the completion of polymerization.

The dimerization competes with monomer addition to monomeric radical-anions,

$$M^{\bar{}} + M \rightarrow \cdot M \cdot M^- , \quad k_a .$$

Most likely k_a is smaller than the propagation constant, i.e. presumably smaller than $<250 \ M^{-1}s^{-1}$. However, since the concentration of the monomer is at least 10^4 times larger than that of the monomeric radical-anions, the addition competes efficiently with the dimerizaton. Nevertheless, the resulting dimeric radical-anions, $\cdot M \cdot M^-$, play no role in the polymerization because their diffusion controlled disproportionation (rate constant $\sim 10^{10} \ M^{-1}s^{-1}$) destroys them as soon as formed. Hence, radical propagation is imperceptible in such systems.

It is appropriate to describe at this junction a technique leading to the determination of the dimerization rate of radical-anions derived from the monomers[96]. Flash-photolysis of $\sim 10^{-6}$ M THF solution of the potassium salt of the dimeric-dianions of α-methyl styrene, $K^+, \bar{}\alpha\alpha^-, K^+$, results in their photodissociation into α-methyl styrene radical-anions:

$$K^+, \bar{}\alpha\alpha^-, K^+ \xrightarrow{h\nu} 2\alpha^{\bar{}}, K^+ .$$

In the dark period following a flash, the latter dimerize and regenerate the original dimers. The rate of dimerization is monitored by the growth of the absorbance at 340 nm (λ_{max} of the dimer) or by the decay of absorbance at 400 or 600 nm caused by the disappearance of the transient. This is shown by Fig. 16. The isosbestic point at 390 nm manifests the stoichiometric and direct conversion of the transient into the dimer; no intermediates could be involved. Eventually, the system returns to its initial state, demonstrating the absence of any side reactions.

---

* This relaxation time is given by $\tau = 1/\{k_f[M] + k_b[A]\}$, k_f and k_b denoting the forward and backward rate constants of the electron transfer. Thus, τ is at the most 1 μs because either forward or backward electron transfer is exothermic and then its rate is diffusion controlled or only slightly slower

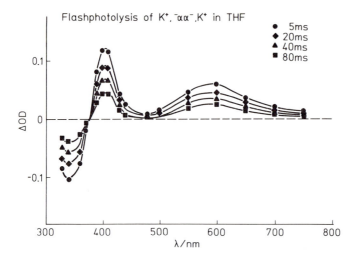

Fig. 16. The difference spectra of flash-photolysed THF solution of K^+, $^-\alpha\alpha^-$, K^+ recorded 5 ms, 20 ms, 40 ms, and 80 ms after a flash

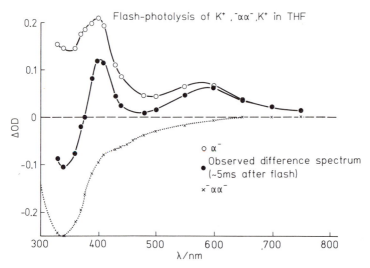

Fig. 17. ●—●—● The difference spectrum of THF solution of K^+, $^-\alpha\alpha^-$, K^+ recorded 5 ms after a flash
×---×---× The constructed absorption spectrum of K^+, $^-\alpha\alpha^-$, K^+ at concentration corresponding to that of the photolysed dimer
○–○–○ The sum of the spectra (a) and (b) giving the absorption spectrum of the transient, α^-, K^+ (radical anion of α-methyl styrene)

The spectrum of the transient, shown in Fig. 17, agrees with that reported for $\alpha^{\overline{}}$ radical-anion obtained by pulse-radiolysis of solutions of α-methyl styrene[498]. This justifies the identification of the transient with α^-, K^+.

Plots of $1/\Delta od$ vs. time, shown in Fig. 18, ($\Delta od = od_\infty - od_t$), are linear demonstrating the second order character of the reaction; their slopes give therefore the dimerization

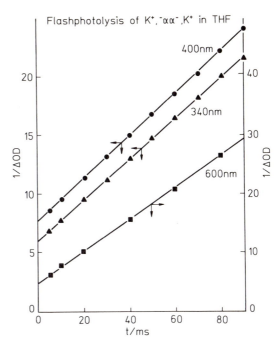

Fig. 18. Plot of 1/Δ (opt. density) vs. time for the flash-photolysed solution of $K^+, ^-\alpha\alpha^-, K^+$ in THF. Recorded at 400 nm, 340 nm, and 600 nm

constant divided by the respective molar absorbance $\Delta\varepsilon$. From such data the dimerization constant was calculated to be $1.0 \times 10^7 \, M^{-1}s^{-1}$.

Similar experiments were performed with the sodium salt ($\sim 10^{-6}$ M), but in this system it was necessary to add ~ 100-fold excess of α-methyl styrene to ensure the linearity of the plot 1/Δod vs. time. The addition of the monomer leads to a negligible conversion of $Na^+, ^-\alpha\alpha^-, Na^+$ into $Na^+, ^-\alpha\alpha\alpha^-, Na^+$ (equilibrium constant of this reaction is $\sim 300 \, M^{-1}$); however, the excess of the monomer was needed to depress the dissociation of $\alpha^{\overline{\cdot}}, Na^+$ into $\alpha + e^-, Na^+$ (see later for the discussion of this point). The spectrum of the transient was virtually identical with that observed in the K^+ system, confirming its assignment to $\alpha^{\overline{\cdot}}, Na^+$. From the linear plot of 1/Δod versus time, the respective dimerization constant was calculated at $0.2 \times 10^7 \, M^{-1}s^{-1}$, substantially lower than that of the potassium salt.

Flash-photolysis of the potassium or cesium salts of the dimeric-dianions of 1,1-diphenyl ethylene, $Cat^+, ^-DD^-, Cat^{+},$[82], resembles in its behavior the photolysis of $K^+, ^-\alpha\alpha^-, K^+$. Since the addition of 1,1-diphenyl ethylene, D, to the dimer is prevented by steric hindrance, the photolysis could be performed in the presence or absence of the unreduced hydrocarbon. In either case the plots 1/Δod versus time are linear, their slopes being unaffected by the addition of D or by its concentration. The dimerization constant of $D^{\overline{\cdot}}, K^+$ was found to be $1 \times 10^9 \, M^{-1}s^{-1}$ and of $D^{\overline{\cdot}}, Cs^+ - 3 \times 10^9 \, M^{-1}s^{-1}$. The much higher dimerization constants of $D^{\overline{\cdot}}$ salts, compared to that of $\alpha^{\overline{\cdot}}, K^+$, are puzzling. They probably arise from a presumably looser structure of those salts than of the salts of $\alpha^{\overline{\cdot}}$.

Flash-photolysis of the lithium or sodium salts of the 1,1-diphenyl ethylene dimer is more complex. The simple behavior resulting in linear plots of 1/Δod versus time are observed only in the presence of a sufficient excess of D. However, at lower concentration of this hydrocarbon, the plots are curved and the curvature increases with decreasing concentration of D. Analysis of the kinetic data led to the suggestion that D^-, Li^+ and D^-, Na^+ are partially dissociated into D and e^-, Li^+ or e^-, Na^+ and that the degree of dissociation of the lithium salt is much greater than of sodium. The assumption of the electron release and its capture occurring simultaneously with the dimerization accounts quantitatively for the experimental findings. From the linear plots obtained at sufficient excess of D, the dimerization constants of D^-, Li^+ and D^-, Na^+ were found to be 1.2×10^8 and 3.5×10^8 $M^{-1}s^{-1}$, respectively. The importance of counter-ions and the character of pairing is clearly revealed by these investigations and this needs stressing.

Not much is known about the rates of dissociation of the dimeric-dianions. An approach yielding the required data was described by Asami and Szwarc[97]. Dimeric-dianions of α-methyl-styrene perdeuterated in the phenyl rings, i.e. $^-\alpha_{5D}\alpha_{5D}^-$ were prepared. Their potassium salt mixed with THF solution of the salt of the ordinary protic dimers slowly reacts yielding mixed dimers, $K^+, ^-\alpha_{5D}\alpha^-, K^+$. Such a reaction is accounted for by the following mechanism:

$$K^+, ^-\alpha\alpha^-, K^+ \rightleftharpoons 2\alpha^-, K^+ ; \qquad K^+, ^-\alpha_{5D}\alpha^-_{5D}, K^+ \rightleftharpoons 2\alpha^-_{5D}, K^+$$

$$\alpha^-, K^+ + \alpha^-_{5D}, K^+ \rightarrow K^+, ^-\alpha\alpha^-_{5D}, K^+ ,$$

and its initial rate in a 50:50 mixture gives ½ the dissociation rate of the dimers. The progress of exchange was followed by withdrawing at desired times aliquots of the mixture, protonating the dimers, and determining with the aid of a mass-spectrometer the composition of the resulting hydrocarbons. By this approach the dissociation constant was determined at 6×10^{-8} s^{-1} at 25 °C.

Knowledge of the dimerization and dissociation constants leads to the respective equilibrium constant of 6×10^{-15} M^{-1}. Assuming a value of 15 eu for the entropy of dissociation, we calculate the heat of dissociation $\Delta H \sim 25$ kcal/mol.

Another approach to the dissociation study was outlined by Spach et al.[98] who investigated the exchange between the radioactive, ^{14}C labeled 1,1-diphenyl ethylene and its dimeric-dianion, $Na^+, ^-DD^-, Na^+$. The exchange demands the dissociation and eventual regeneration of the dimers, and the rate of dissociation determines the rate of incorporation of the radioactivity. Two routes are available for the dissociation, a direct one,

$$Na^+, ^-DD^-, Na^+ \xrightarrow{k_1} 2D^-, Na^+, \qquad K_{diss} ,$$

and an indirect one involving an electron-transfer process,

$$Na^+, ^-DD^-, Na^+ + D \rightleftharpoons Na^+, ^-DD\cdot + D^-, Na^+, \qquad K_{ex}$$

followed by

$$Na^+, ^-DD\cdot \underset{k_{-2}}{\overset{k_2}{\rightleftharpoons}} D + D^-, Na^+ .$$

Participation of both processes in the exchange leads to the following rate expression:

(Rate of exchange)$/[Na^+, \bar{}DD^-, Na^+] = k_1 + k_{-2}K_{diss}^{1/2}[D]/[Na^+, \bar{}DD^-, Na^+]^{1/2}$,

which is justified by the condition $k_2[Na^+, \bar{}DD^-, Na^+] = k_{-2}[D][D\bar{}, Na^+]$ describing the equilibrium state of the system. Indeed, the plot of (Rate of exchange)/$[Na^+, \bar{}DD^-, Na^+]$ versus $[D]/[Na^+, \bar{}DD^-, Na^+]^{1/2}$ is linear. Its intercept is too small to reliably determine k_1 and provides only its upper limit, i.e., $k_1 \leq 10^{-7} s^{-1}$. The slopes give at 25 °C

$$k_{-2}K_1^{1/2} = 35 \times 10^{-6} M^{-1/2}s^{-1} .$$

Assuming a value of $\sim 10^{-14} M^{-1}$ for K_{ex} we find $k_{-2} \sim 300 M^{-1}s^{-1}$. Hence, the association $D + D\bar{}, Na^+$ is relatively slow; even at 10^5-fold excess of the monomer over the initiator it contributes only 50% to the dimer formation. Provided that the above findings could be generalized for other monomers, the reaction

$$M\bar{}, Cat^+ + M \rightarrow \cdot M \cdot M^-, Cat^+$$

is of little importance in the initiation by electron-transfer. The $\cdot MM^-, Cat^+$ dimers left at the end of the initiation process probably are rapidly destroyed by a disproportionation involving electron-transfer,

$$2 \cdot MM^-, Cat^+ \rightarrow Cat^+, \bar{}MM^-, Cat^+ + 2M ,$$

a reaction more plausible than their coupling into a tetramer.

In our discussion we tacitly assumed that the vinyl or vinylidene radical-anions dimerize by the tail-to-tail coupling. This mode of coupling is thermodynamically the most probable; whether other modes occur in some systems is still unknown. However, carboxylation of the dimers derived from styrene, α-methyl styrene and 1,1-diphenyl ethylene, conclusively proved that the 1,4-dicarboxylic acids derived from the tail-to-tail coupling are the only detectable products of the reaction[99].

Exceptional products of dimerization are formed by the sterically hindered radical-anions. For example, protonation of the dimeric dianions of 1-Phenyl-2-t-butyl acetylene yielded the following hydrocarbons[100]

and

This mode of dimerization is analogous to that operating in the dimerization of the sterically hindered tri-phenyl-methyl radicals[101].

Radical-anions derived from aromatic hydrocarbons of higher electron-affinity are rather poor and complex initiators. For example, anthracene radical-anions do not initiate polymerization of styrene. Moreover, the parent unreduced aromatics react with carbanions in a two-fold way. They oxidize carbanions to the respective radicals or alternatively they combine with them, yielding relatively unreacted adducts which do not propagate. Addition of anthracene (An) to living polystyrene ($\sim$ S$^-$) serves as an example[102],

$$\sim S^- + An \rightarrow \sim SAn^- .$$

On addition of a stoichiometric amount of anthracene the characteristic spectrum of living polystyrene is replaced by that of the adduct, ($\sim$ SAn$^-$). The rate of styrene addition, i.e., propagation, becomes very slow and analysis of the kinetic data indicates that the observed reaction arises from the addition to a minute fraction of the active, living polymers, $\sim$ S$^-$, which are in equilibrium with the non-active, dormant polymers, $\sim$ SAn$^-$. Hence, the rate of propagation decreases inversially proportionally with the increasing concentration of the anthracene excess[102].

Electron-transfer from carbanions to aromatic hydrocarbons like anthracene has been known from the beginning of this century. For example, the oxidation of triphenyl carbanions, Ph$_3$C$^-$, by benzophenone was described by Schlenk in 1916. Sufficiently powerful acceptors such as aromatic nitro-compounds may acquire electrons from poor donors such as alkoxides or thiolates. Butyllithium is oxidized by sterically hindered olefines, e.g., its reaction with 1,2,3,4-tetra-phenyl butadiene leads to electron transfer instead of addition.

Two kinetic studies of electron-transfer from carbanions to aromatic acceptors will be discussed. The reversible interaction of anthracene, An, with the dimeric-dianions of 1,1-diphenyl ethylene, Na$^+$, $^-$DD$^-$, Na$^+$, leads to an equilibrium mixture[103],

$$Na^+, {}^-DD^-, Na^+ + 2\,An \rightleftharpoons 2\,An^{-\cdot}, Na^+ + 2\,D .$$

Kinetics of the forward and backward reactions reveal that the decomposition of the dimeric radical-anion, viz.

$$Na^+, {}^-DD \cdot \xrightarrow{k_2} D^{-}, Na^+ + D ,$$

followed by a rapid electron-transfer,

$$D^{-}, Na^+ + An \rightarrow D + An^{-}, Na^+ ,$$

is the rate determining step of the forward process. The concentration of Na$^+$, $^-$DD$\cdot$ is determined by the rapidly established equilibrium,

$$Na^+, {}^-DD^-, Na^+ + An \rightleftharpoons Na^+, {}^-DD \cdot + An^{-}, Na^+, \quad K_{An} .$$

Thus the spectrophotometrically monitored rate of disappearance of Na$^+$, $^-$DD$^-$, Na$^+$, equal to ½ the rate of formation of A^{-}, Na$^+$, is given by

$$-d[Na^+, {}^-DD^-, Na^+]/dt = (1/2)d[A^{-}, Na^+]/dt =$$

$$k_2 K_{An}[Na^+, {}^-DD^-, Na^+][An]/[An^{-}, Na^+] .$$

Similar observations led to analogous expressions for the reactions involving 1,2-dimethyl-anthracene, DMA, or pyrene, π, as the electron acceptors[104]. The combined constants, $k_2 K_{DMA}$ and $k_2 K_\pi$, derived from these studies

$$Na^+, {}^-DD^-, Na^+ + DMA \rightleftharpoons Na^+, {}^-DD\cdot + DMA^{\overline{\cdot}}, Na^+ , \qquad\qquad K_{DMA}$$

$$Na^+, {}^-DD^-, Na^+ + \pi \rightleftharpoons Na^+, {}^-DD\cdot + \pi^{\overline{\cdot}}, Na^+ , \qquad\qquad K_\pi .$$

confirm the proposed mechanism by showing the agreement between the ratios $k_2 K_{DMA}/k_2 K_{An}$ or $k_2 K_\pi/k_2 K_{An}$ and the respective electron-transfer constants, i.e.,

$$DMA^{\overline{\cdot}}, Na^+ + An \rightleftharpoons DMA + An^{\overline{\cdot}}, Na^+ \qquad\qquad K_{DMA, An}$$

and

$$\pi^{\overline{\cdot}}, Na^+ + An \rightleftharpoons \pi + An^{\overline{\cdot}}, Na^+ , \qquad\qquad K_{\pi, An}$$

the latter constants being determined by potentiometric methods.

No intermediate adduct was seen in the above reaction. However, a similar electron-transfer from the dimeric-dianions of α-methyl styrene, α, to anthracene, An, namely

$$K^+, {}^-\alpha\alpha^-, K^+ + 2\,An \rightarrow 2\,An^{\overline{\cdot}}, K^+ + 2\,\alpha ,$$

reveals the formation of an intermediate adduct[105]. Addition of anthracene to the dimer yields rapidly, within less than a second, an adduct, presence of which is revealed by the absorbance at 451 nm. The intensity of this absorbance increases with increasing concentration of the added anthracene, as shown in Fig. 19, and reaches a plateau value when the ratio $[An]/[K^+, {}^-\alpha\alpha^-, K^+]$ attains the value of 2. It appears therefore that both mono- and di-adducts are formed, i.e., $K^+, {}^-\alpha\alpha An^-, K^+$ and $K^+, {}^-An\alpha\alpha An^-, K^+$. The mono-adduct decomposes within ~ 10 s yielding $An^{\overline{\cdot}} K^+$ as revealed by its characteristic absorbance at 720 nm. The initial rate of this decomposition increases with the ratio $[An]/[K^+, {}^-\alpha\alpha^-, K^+]$, but it reaches a maximum at its unity value, thereafter decreases and eventually comes to a low plateau value as the ratio exceeds 2. These observations, shown in Fig. 20, imply a relatively fast decomposition of the mono-adduct,

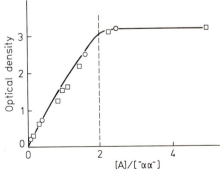

Fig. 19. Optical density at 451 nm (λ_{max} of the transient) for increasing ratio [Anthracene]/[K$^+$, $^-\alpha\alpha^-$, K$^+$], recorded 2 s after mixing THF solution of K$^+$, $^-\alpha\alpha^-$, K$^+$ with anthracene. [K$^+$, $^-\alpha\alpha^-$, K$^+$] = const

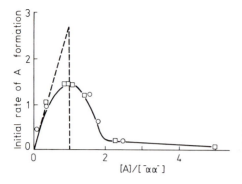

Fig. 20. Initial rate of formation of anthracene$^+$ radical-anions in THF solution of $K^+, ^-\alpha\alpha^-, K^+$ with variable proportion of anthracene. $[K^+, ^-\alpha\alpha^-, K^+]$ = const. Monitored by the absorbance at 720 nm (λ_{max} of anthracene$^+$)

$K^+, ^-\alpha\alpha A^-, K^+$; but very slow, of the di-adduct, $K^+, ^-A\alpha\alpha A^-, K^+$. The kinetic data led to the following rate of the mono-adduct decomposition,

$$d[An^{\overline{\cdot}}, K^+]/dt = const\,[K^+, ^-\alpha\alpha An^-, K^+]/[A^{\overline{\cdot}}, K^+]\,,$$

accounted for by the mechanism

$K^+, ^-\alpha\alpha An^-, K^+ \rightleftharpoons K^+, ^-\alpha\alpha \bullet + An^{\overline{\cdot}}, K^+$,   rapidly established equilibrium,

$K^+, ^-\alpha\alpha \bullet \to \alpha^{\overline{\cdot}}, K^+ + \alpha$,   rate determining step,

followed by the rapid electron-transfer or coupling step

$K^+, ^-\alpha\alpha \bullet + \alpha^{\overline{\cdot}}, K^+ \to K^+, ^-\alpha\alpha^-, K^+ + \alpha$ or $K^+, ^-\alpha\alpha\alpha^-, K^+$.

At constant initial concentration of the dimer, the initial rate of decomposition is given by const. $2p(1-p)$ where $p = [An]_0/2[K^+, ^-\alpha\alpha^-, K^+]_0$. This expression requires equal probability of anthracene addition to either end of the dimer, independent of the fate of the other end – whether associated or not with another anthracene molecule. A similar behavior was observed in a system with pyrene replacing anthracene as the electron acceptor.

The above examples show the complexity of the systems involving radical-anions derived from compounds of higher electron-affinity. It is not surprising, therefore, that benzophenone ketyl and other similar compounds do not initiate styrene polymerization, although they initiate polymerization of acrylonitrile or methyl-methacrylate. On the other hand, the monomeric dianions of benzophenone initiate polymerization of styrene as well as of other monomers, but not of vinyl chloride or acetate. Mechanisms of these initations were not investigated and presumably are complex.

Electron-transfer to monomer is not the only mode of initiation by radical-anions, although it is unique for non-polar monomers. With some polar monomers, especially cyclic ones, the initiation resembles protonation[106]. For example, the reaction of sodium naphthalenide with ethylene oxide follows the route[107].

$$\text{naphthalene}^{\overline{\cdot}} + \underset{\text{O}}{\overset{\text{CH}_2—\text{CH}_2}{\triangle}} \longrightarrow \text{naphthalene–}\overset{\text{H}\;\;\text{CH}_2\text{CH}_2\text{O}^-}{\diagup}$$

Its kinetics was investigatged and, the bimolecular rate constant was found to be $\sim 1\,M^{-1}s^{-1}$ [52, 161].

The reduction of the adduct by naphthalenide followed by the addition of another molecule of epoxide yields the para or ortho diaduct endowed with alkoxide groups,

The subsequent propagation is due to alkoxide ions. A similar process was reported for the initiation of cyclic-tetra-dimethylsiloxane by naphthalenide[108].

The evidence for the proposed mechanism is two-fold: (a) the presence of aromatic moiety in the resulting poly-glycol and (b) the quantitative analysis of the solution left after precipitation of the polymer demonstrating that only one-half of the utilized naphthalenide was converted into naphthalene*.

Another variant of initiation by naphthalenide was proposed by Sigwalt[109], who studied the anionic polymerization of propylene sulphide. Formation of propylene in the course of this reaction led him to propose the following mechanism:

sodium naphthalenide + propylene sulphide $\rightarrow$
naphthalene + NaS$\cdot$ + propylene .

NaS$\cdot$ dimerize and the formed sodium disulphide, NaSSNa, initiates the propagation. The presence of a S–S bond in the resulting polymer was confirmed by reducing them with sodium fluorenyl to –SH-bonds[463]

III.6. Initiation by Anions or their Ion-Pairs

Initiation of anionic polymerization by anions is conceptually similar to propagation. Indeed, every living polymer is an initiator of its own monomer polymerization. However, not every anion is capable of initiating polymerization of a specified monomer. It is tempting, therefore, to assign to each initiator a value measuring its nucleophilic power, and in a similar fashion, a value could be assigned to each monomer gauging its acceptance power. The initiators could be arrayed then in a series of ascending nucleophilic powers, with an analogous series of electrophilic powers constructed for the monomers. Such an approach implies that all the initiators of nucleophilic power exceeding some limiting value should initiate polymerization of a specified monomer, as well as that all the monomers of acceptance power greater than a limiting one could be polymerized by a specified initiator.

---

* It has been claimed that at very low concentrations of the naphthalenide, polymers with only one growing group are formed[161]. Hydrogen abstraction from solvent was proposed as an explanation. However, terminating impurities become significant at very low concentrations of initiators and their action may account for the observations

Unfortunately, the problem is more complex than outlined above. The sequence of initiators varies with the nature of the polymerized monomer, steric effects often create "exceptions", and the state of aggregation of the initiating species, whether it acts as a free ion or an ion-pair, etc., is an important factors affecting its reactivity. Hence, the most one could hope for is to construct ordered series for related initiators, gauging their ability to initiate polymerization of a specified family of monomers of a similar nature.

A few examples are in place. Carbanions are good initiators of polymerization of many vinyl and diene monomers. Their reactivities decrease along the series: primary, secondary, tertiary. Thus, benzyl carbanion is a poor initiator of styrene polymerization, whereas the oligomeric α-methyl styrene anions or cumyl carbanions are very efficient[110].

Salts of fluorenyl carbanions or of derivatives of diphenyl-methyl carbanions do not initiate polymerisation of styrene but are efficient and clean initiatiors of methyl methacrylate polymerisation.

Free anions are often more powerful initiators than their ion-pairs, especially for non-polar monomers, although a reverse order of reactivities is observed for some cyclic monomers. Polymerization of styrene in liquid ammonia is initiated by free amide ions but not by their potassium ion-pairs[111]. The enormous influence of the aggregation state of an initiating species, as well as of its surrounding, is dramatically revealed by the startling effects of crown ethers and kryptates on their reactivity. Kryptates are tricyclic compounds, synthesized by Lehn[119], that could encapsulate ions and greatly affect their behavior. This was shown by the numerous examples reported by Lehn (see, e.g. ref. 119). It is well known that polymerization of styrene is not initiated by alkoxides; even in

2,2,2 Cryptand   18-Crown-6
(C$_{222}$)   (18-C-6)

liquid ammonia where some dissociation of alkoxides into free RO$^-$ ions does take place[120] polymerization is not observable. Alkoxides are hardly soluble in hydrocarbons, but their solubility is greatly increased on addition of kryptates[117, 118, 121], e.g. about 10^{-4} M solution of kryptated sodium tertioamylate in benzene could be prepared, and it converts styrene into its oligomers[116]. Unfortunately, some side reactions interfere with the initiation*.

Some comments are helpful. Kryptates encapsulate alkali or akaline-earth cations and convert tight ion-pairs, or some still larger aggregates, into loose pairs. This transformation decreases the lattice energy of the salt. Moreover, the interaction of kryptated cations with hydrocarbon solvents is improved because it is the external, hydrophobic shell of the complex that contacts the surroundings. The combined effect of both factors makes the complexes more soluble in hydrocarbons than the uncomplexed salts.

---

* Although quantitative conversion was claimed in Ref. 116, re-examination of the system showed a low yield. Presumably, the alkoxides react with the kryptate (Private communication of Prof. F. Schué.)

Interaction of anions with hydrophobic reagents like styrene is hindered in polar solvents. The anion is surrounded by strongly interacting polar molecules, and the hydrophobic reagent has to displace some of them to allow for a reaction. This is an unfavorable step. On the other hand, in hydrocarbon media the opportunity for a styrene molecule to contact the anion is greatly improved. Moreover, the partial separation of ions in a loose pair facilitates the transfer of the negative charge to the reagent allowing for the reaction

$$\left(Na^+\right),\ \bar{O}R + CH_2{:}CHPh \longrightarrow ROCH_2{\cdot}\bar{C}HPh,\ \left(Na^+\right)$$

Thus, the apparent paradox is explained. The free $R\bar{O}$ anion in a hydrophilic solvent does not react with styrene, whereas the reaction might take place with its loose kryptated ion-pair in a hydrophobic milieu.

By the same token, kryptated potassium hydroxide initiates polymerization of hexamethyl-cyclotrisiloxane in benzene but not in polar solvent. About 10^{-4} M solution of kryptated KOH converts in about five minutes 60% of $5 \cdot 10^{-2}$ M siloxane into its polymer. Other examples are given in Ref. 116.

Numerous complexes of amides and alkoxides involving Li^+, Na^+ or K^+ as the cations were described by Canber, Lecolier et al.[443], who stressed their versatility in affecting the rate and stereochemistry of polymerization.

Kinetics of initiation of epoxide and thiirane polymerizations by salts of anions derived from carbazole, fluorene and their derivatives was intensively studied by Sigwalt and Boileau[112, 113] and by Hogen-Esch[114]. The latter worker investigated the cleavage of ethylene oxide by fluoradenyl salts. This nucleophilic reagent exists in ethylene oxide solution in the form of loose, Li^+, Na^+ or K^+ ion-pairs, whereas the Cs^+ salt forms in this medium only tight pairs. The above statement is based on spectroscopic observations. Interestingly, ethylene oxide appears to be a better agent for the conversion of tight into loose ion-pairs than oxetane or tetrahydrofuran, contrary to the well documented basicity sequence of cyclic oxides[115]. The reversal of the order probably is caused by the smallness of the oxide, allowing a larger number of its molecules to coordinate with each cation. A similar explanation was offered by Kebarle[467] to account for the reversal of solvating power of methanol and water in the gas phase and in solution.

In the ethylene oxide system the free ions seem to be less reactive than the ion-pairs. Significantly, the addition of an appropriate tetraphenyl boride salt that depresses the dissociation of ion-pairs into free ions speeds up the cleavage, while the addition of crown ethers or other powerful solvating agents slows it down. The association of the developing alkoxide ion with the cation of a pair facilitates the reaction. Since the coordination of a cation with a crown ether makes it inaccessible, or at least less accessible, for the interaction with the oxide, the cleavage becomes slower in the presence of a crown. In fact, it appears that a molecule of ethylene oxide coordinated with a cation is more reactive than a non-associated one, this accounts probably for the observed enhancement of the reaction resulting from the addition of an appropriate boride salt.

The cleavage of ethylene oxide by alkali salts of 9-methyl-fluorenyl anion was investigated in several ethereal solvents by Sigwalt and Boileau[112]. At a salt concentration lower than 10^{-4} M, the reaction is first order in the oxide and in the salt. However, at higher concentrations of the salt some aggregation complicates the kinetics. The contact ion-pairs were found to be more reactive than the kryptated pairs or free anions, like in the Hogen-Esch studies. For example, the bimolecular rate constant of the cleavage, determined for tetrahydropyrane solution at $-30\,°C$, is $1.5 \cdot 10^{-2}$ $M^{-1}s^{-1}$ for the tight sodium salt, but only $0.5 \cdot 10^{-6}$ $M^{-1}s^{-1}$ for the kryptated salt. Interestingly, the rate constant is about three times lower in tetrahydrofuran than in tetrahydropyrane, presumably because ethylene oxide has to compete with the solvent for an association site on the cation, and the better the solvating power of the solvent the more difficult becomes the competition.

The carbazyl salts, excellent initiators of oxirane polymerization[112], behave differently than the fluorenyl or fluoradenyl salts; their free ions are more reactive cleaving agents than their ion-pairs. The effects arising from the interaction of cation with the developing anion were stressed in our discussion; however, as was pointed out by Sigwalt[113], another facet of the problem also deserves consideration. The reactivity of anions, whether in the free form or paired with cations, should increase with increasing density of negative charge on its reaction center. On dissociation of a carbazyl ion-pair the nitrogen atom, its reactive center, retains the negative charge to a greater extent than the 9-carbon atom of a dissociating fluorenyl ion-pair. In the latter anion charge delocalization is greater than in the former. Sigwalt argued that this decrease of the negative charge on the $9 - C$ of the flurenyl anion resulting from the dissociation of its ion-pair, makes the free anion less reactive than the ion-pair. Due to the lesser degree of delocalization in the carbazyl anion, the reactivity of its free ion is reduced only slightly and the conventional order – free ion more reactive than its ion-pair – is retained in that system.

To substantiate his thesis, Sigwalt compared the initiating capacity of the free ions and ion-pairs of carbazyl with those of dibenzocarbazyl. Kinetic studies of the epoxide cleavage in tetrahydrofuran at $20\,°C$ led to the following bimolecular rate constants:

Cbz^-, K^+ $\qquad$ $10^4 \cdot k_\pm$ $= 3.5$ $M^{-1}s^{-1}$; $10^4 k_- \sim 10 \cdot M^{-1}s^{-1}$

$DibenzoCbz^-, K^+$ $\qquad$ $10^4 k_\pm$ $= 1$ $M^{-1}s^{-1}$; $10^4 k_- \sim 0.4$ $M^{-1}s^{-1}$

(The rate constants k_- were derived from studies of the kryptated salts.)

$\quad$ Cbz^- = Carbazyl$^-$ $\qquad\qquad$ $^-$Dibenzo-carbazyl$^-$

The greater delocalization of the negative charge in the dibenzo derivative when compared with the carbazyl salt makes its free ion as well as its ion-pair less reactive than the respective species of the carbazyl salt and, significantly, the order of reactivities of the free ions and ion-pairs is reversed. This observation supports Sigwalt's suggestion.

Further discussion of the ion-pairing effect is continued in the section dealing with the propagation steps. However, one point deserves stressing. The usefulness of an initiator is determined by two factors. Firstly, how readily it reacts with a monomer in producing its respective anion, and the preceding discussion was concerned with this aspect of the problem. Secondly, the anion derived from a monomer may or may not interact with the cation introduced by the initiator. Hence, such cations that produce the most reactive propagating species are the most desired. Here the effect of kryptates is of a great importance since in many propagation processes the kryptated ion-pairs are more reactive than the tight pairs.

Other examples of simple initiation of anionic polymerization by anions are provided by the work of Butler[212], who described the preparation of living poly-nitrostyrene and its derivative with alkoxides as the initiators; by studies of Zilkha[213], who initiated polymerization of 4-vinyl pyridine by potassium methoxide in dimethyl sulphoxide; and by the extensive studies of Shashoua[214], who initiated polymerization of isocyanates by potassium cyanide in dimethyl-formamide at $-60\,^\circ$C. The latter reaction yields Nylon-1 and is extremely fast even at this low temperature. Strangely enough, polymerization of cyclohexyl-isocyanate, $C_6H_{11} \cdot CNO$, initiated by sodium naphthalenide in tetrahydrofuran, was inefficient and yielded only a low molecular weight polymer[218]. A review covering polymerization of various isocyanates was published recently[219].

A few practical remarks are in place. Most of the initiators have to be freshly prepared before being used since a variety of side reactions destroys, or at least reduces, their usefulness[110, 122].

Commercial products are often unsatisfactory in kinetic studies. Preparation of some commonly used initiators is given elsewhere[141]. Benzyl sodium is prepared from mercury dibenzyl and metallic sodium[123]; a similar technique yields barium and strontium salts. Fluorenyl lithium, sodium etc.*, are conveniently prepared from fluorene and a suitable organo-metallic compound, e.g.

$$\text{CH}_2 + \text{EtLi} \longrightarrow \text{CH}^-, \text{Li}^+ + \text{C}_2\text{H}_6$$

Direct reaction of fluorene with alkali metals is not recommended. It proceeds via radical-anions and yields some side products.

In a similar fashion one prepares 1,2-diphenyl-hexyl lithium by reacting n-butyl lithium with 1,1-diphenyl ethylene. It is a relatively slow initiator for hydrocarbon monomers but convenient for acrylates and methacrylates. Its use precludes side reactions that occur with other initiators. Preparation of cumyl potassium was described by Ziegler[124] and the same method yields the respective cesium salt. This initiator is stable and shows no change on storage at ambient temperature. After completion of the preparation, it is advantageous to chill the solution to $-70\,^\circ$C and after a few hours to filter or decant it from the precipitated methoxide.

---

* The fluorenyl moiety on the end of a polymer may act as a chain transfer agent. Its acidic 9-proton may terminate a growing polymer yielding a fluorenyl anion substituted in 9 position by a polymer chain

III.7. Lithium Alkyls

Lithium alkyls are important and frequently used initiators of anionic polymerization. To understand their behavior it is advisable to review their peculiar features. Alkyl lithiums are aggregated even in the most dilute solutions or in rarified vapors. The aggregates belong to the class of electron-deficient compounds – those having a greater number of bonds, i.e., the nearest neighbor atom-atom connections, than the number of available pairs of valence electrons[125]. Their degree of aggregation varies with the nature of alkyl groups and solvent. As a general rule, the degree of aggregation decreases with the bulkiness of the alkyl group and the increasing polarity of the solvent.

Colligative properties of alkyl lithium solutions show n-butyl and ethyl lithium to be hexameric in hydrocarbons but tetrameric in diethyl ether[126, 127]; t-butyl lithium and sec-butyl lithium are tetrameric both in hydrocarbons and in ethers[128, 156]; methyl lithium, the only alkyl lithium insoluble in hydrocarbons, is tetrameric in diethyl ether[143, 150] etc. The measured degrees of association in hydrocarbons are not affected by the concentration of RLi, implying that in solutions of a lithium alkyl a single kind of aggregate dominates, even at high dilutions*.

In the pure state n-butyl-, sec-butyl-, and iso-butyl lithiums are liquids at ambient temperatures, whereas methyl-, ethyl- and t-butyl lithiums are crystallinic solids. X-ray studies revealed that the crystals of methyl lithium are built of tetramers[132, 135]; a similar structure was deduced for t-butyl lithium[128, 133]; whereas a layer structure, still built of tetrameric units, was proposed for ethyl lithium[136]. The inability of n-propyl-, n-butyl-, iso-butyl-, sec-butyl-, and amyl-lithium to crystallize seems to be caused by steric factors.

The high degree of aggregation of alkyl lithiums is responsible for their low volatility, e.g. the vapor pressure of n-butyl lithium at 60 °C is only $4 \cdot 10^{-4}$ tor[129]. The aggregation is also responsible for the solubility of alkyl lithiums in hydrocarbons. The partially ionic C–Li bonds form an inner core surrounded by the hydrocarbon groups, thus imparting a hydrophobic property to the aggregates.

The nature of bonding in the alkyl lithium aggregates was established by the extensive studies of Brown[134, 137] and by the X-ray investigations of Weiss[132, 135]. The latter worker showed the tetramers of methyl lithium to be composed of four lithium atoms forming a tight, tetrahedrally shaped core with the methyl groups placed above the respective faces of the tetrahedron, as shown in Fig. 21. The Li–Li distance is 2.5 Å, being shorter than the Li–Li bond lengths (2.67 Å) in the Li_2 molecule. The sharpness of the Li–C–Li angles is indicative of electron-deficient bridging, i.e., each Li atom seems to be associated with three carbon atoms of the alkyl groups, the former contributing three 2p orbitals while each carbon atom provides a single sp^3 orbital[128, 133]. Their overlap results in molecular orbitals joining four centers but accommodating only four valence electrons. The average bond energy for each bonding electron pair is calculated to be 82 kcal/mol on the basis of the available thermochemical data[134].

A similar structure applies to the tetrameric ethyl lithium and t-butyl lithium, whereas a more complex structure, involving Li–C bonds as well as H bridges, was proposed for the hexamers[142]. This is shown in Fig. 22.

---

* Some dissociation of aggregates is observed in highly rarified vapor. For example, vapor of ethyl lithium at low pressure is composed of a mixture of tetramers and hexamers[130]

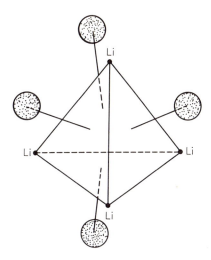

Fig. 21. Tetrahedral structure of the methyl-lithium tetramer. The gray balls represent the methyl groups. Each methyl group interacts with three Li atoms

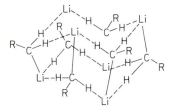

Fig. 22. Structure of the hexameric *n*-butyl lithium. The bonding involves Li-C as well as Li-H bonds

The tightness of the Li core is manifested by the appearance of $Li_4R_3^+$ ions in the mass-spectra of tetramers and $Li_6R_5^+$ and $Li_6R_3^+$ ions in the mass-spectrum of the hexameric ethyl-lithium. These fragments, together with $Li_2R_4^+$, are the most abundant ions formed by the impact of electrons accelerated to 75 V[130]. Nevertheless, there is little, if any, bonding between the Li atoms since no ^6Li–^7Li coupling was observed in alkyl lithiums enriched by ^6Li[138].

According to Brown the chemical shift in the ^7Li-NMR is determined by the "local" environment, i.e. by the nature of the three alkyl groups bonded to a lithium atom. This hypothesis was verified through the NMR studies of equilibrated mixtures of the tetrameric methyl- and ethyl lithium in diethyl ether. The *inter*-molecular exchange yields mixed aggregates having four different "local" environments, namely, 3 Me, 2 Me and 1 Et, 1 Me and 2 Et, and 3 Et. Indeed, four lines appear in the ^7Li-NMR at − 80 °C, the chemical shifts of the extreme ones being those observed for the pure methyl lithium and ethyl lithium, respectively[139]. Their intensities depend on the composition of the mixture, being accounted for by the random distribution of the alkyl groups. Had the *intra*-molecular exchange been rapid, five lines should be observed, i.e., 4 Me, 3 Me and 1 Et, 2 Me and 2 Et, 1 Me and 3 Et, and 4 Et. At higher temperatures the lines broaden and eventually collapse into an average one as a result of rapid *inter*-molecular exchange.

Further evidence for the tetrahedral structure and the "local" environment hypothesis is provided by the determination of the ^{13}C–^7Li couplings of methyl lithium enriched by ^{13}C[140]. The "local" environment hypothesis predicts seven lines, as observed, whereas nine lines would appear had the *intra*-molecular exchange been fast.

The NMR studies allow one to investigate the mechanism of *inter*-molecular exchange. The aggregates are in equilibrium with minute amounts of some less aggregated species, e.g. tetramers could be partially dissociated into dimers. On mixing cyclopentane solutions of *t*-butyl lithium, $(LiR_1)_4$, and lithio-methyl-trimethyl silane, $(LiR_2)_4$, a slow exchange takes place[144]. Interestingly, as the original NMR lines of the components decay, a line corresponding to the mixed tetramer $(LiR_1)_2(LiR_2)_2$ appears before the lines corresponding to $(LiR_1)(LiR_2)_3$ or $(LiR_1)_3(LiR_2)$ are observed. This impies a following mechanism of the exchange:

$$(LiR_1)_4 \rightleftharpoons 2\,(LiR_1)_2\,; \quad (LiR_2)_4 \rightleftharpoons 2\,(LiR_2)_2$$

$$(LiR_1)_2 + (LiR_2)_2 \rightleftharpoons (LiR_1)_2(LiR_2)_2\,.$$

These results could not be accounted for by a direct reaction $(LiR_1)_4 + (LiR_2)_4 \rightarrow$ mixed tetramers, or by the dissociation of tetramers into monomers and trimers. However, a contribution of the steps

$$(LiR_1)_2 + (LiR_2)_4 \rightleftharpoons (LiR_1)_2(LiR_2)_2 + (LiR_2)_2$$

or

$$(LiR_2)_2 + (LiR_1)_4 \rightleftharpoons (LiR_2)_2(LiR_1)_2 + (LiR_1)_2$$

can not be excluded. These reactions may be important whenever the concentration of the dimers is exceedingly low since then their recombination becomes slow.

Continuation of these studies allowed Brown to determine the heat of activation, $\Delta H^\dagger$, and the entropy of activation, $\Delta S^\dagger$, of the dissociation of the tetrameric methyl lithium in diethyl ether and the tetrameric *t*-butyl lithium in cyclopentane[145], namely,

$$(MeLi)_4 \rightleftharpoons 2\,(MeLi)_2 \text{ in diethyl ether}\,; \quad \Delta H^\dagger = 11 \text{ kcal/mol}, \Delta S^\dagger = -6 \text{ eu}.$$

$$(t\text{-BuLi})_4 \rightleftharpoons 2\,(t\text{-BuLi})_2 \text{ in cyclopentane}\,; \quad \Delta H^\dagger = 24 \text{ kcal/mol}, \Delta S^\dagger \sim 0\,.$$

The activation energy of the dissociation of $(MeLi)_4$ in diethyl ether is lower than that of (t-BuLi)$_4$ in cyclopentane apparently because the solvation of the transition state of the former aggregates by the ether facilitates their dissociation. The reported values of the entropies of activation support this explanation.

The dissociation of the tetramers into a monomer and a trimer seems unlikely because fragments involving an odd number of Li atoms are hardly seen in the mass-spectra of the aggregated alkyl lithiums. Nevertheless, it is probable that monomeric lithium alkyls are present in solutions of the aggregates and these could be the direct initiators of polymerization of vinyl and diene monomers. Their low concentration might be offset by their very high reactivity*.

Mixed aggregates are also formed between alkyl lithiums and some other compounds. For example, lithium bromide, which is aggregated in diethyl ether[147, 148], reacts with methyl lithium forming mixed associates, presumably $(MeLi, LiBr)$[146]. Interesting com-

---

* However, see Ref. 162 and p.63

plexing takes place between alkyl lithium and various Lewis bases. The complexation may, or may not, lead to fragmentation of the alkyl lithium aggregates. For example, triethyl amine or lithium ethoxide combine with the hexameric ethyl lithium without degrading the hexamers, provided that the concentration of the base is not too high[131, 149, 150]. Brown presumes that the aggregates have vacant "faces" on which the base becomes accommolated, e.g. a hexamer has two vacant "faces" and therefore it associates with two molecules of the amine. Lack of further vacant sites prevents further complexation; however, since new sites are formed on fragmentation of the aggregates, such a fragmentation is facilitated by a higher concentration of the base.

The degree of fragmentation induced by Lewis bases has been determined by a variety of techniques. Methods utilizing colligative properties are the simplest but the results were questioned[151]. Infra-red studies were helpful. For example, the absorption of aggregates at 500 cm^{-1} is affected on replacement of ^7Li by ^6Li; hence, it is likely that this band is associated with C–Li bending and stretching. It disappears as the aggregates dissociate into monomers[152]. The dielectric constant of solutions of alkyl lithiums mixed with a Lewis base provides another diagnostic tool revealing the complexation or fragmentation of the aggregates[150, 153]. Studies of the composition of saturated vapor at equilibrium with alkyl lithium solution provide still another avenue of approach leading to the stoichiometry of the complexation[154].

The fragmentation or complexation of alkyl lithium aggregates has a profound effect upon their ability to initiate anionic polymerization and it greatly affects their rate of initiation. Examples of such effects are discussed in the next section.

III.8. Mechanisms of Initiation of Anionic Polymerization by Alkyl Lithiums

Lithium alkyls are extremely reactive and versatile reagents. They rapidly react with oxygen, carbon dioxide, moisture, alcohols, ethers, ketones, etc. For example,

$$RLi + O_2 \rightarrow ROOLi \xrightarrow{RLi} 2\,ROLi,$$

$$RLi + CO_2 \rightarrow RCOOLi,$$

$$RLi + H_2O \text{ or } R'OH \rightarrow RH + LiOH \text{ or } R'OLi,$$

$$RLi + EtOEt \rightarrow RH + C_2H_4 + EtOLi,$$

$$RLi + \text{(cyclic ether)} \longrightarrow RH + CH_2{:}CHOLi + C_2H_4,$$

$$RLi + R'CO \cdot R'' \rightarrow R \cdot (R')(R'')COLi.$$

Not surprisingly, they readily add to the C=C bonds, especially those of styrene, dienes and their derivatives. These reactions, formally represented as

$$RLi + \overset{|\quad|}{\underset{|\quad|}{C{=}C}} \rightarrow \overset{|\quad|}{\underset{|\quad|}{RC{-}C}} \cdot Li,$$

are the initiation processes of anionic polymerization, e.g.

$$RLi + CH_2:CHPh \rightarrow RCH_2 \cdot {}^-CH(Ph), Li^+ .$$

The rates of those additions are relatively slow compared with the rates of reactions involving H_2O, CO_2 or alcohols. Hence, the presence of these agents prevents the initiation by destroying the initiators. On the other hand, reactions involving ethers, esters, nitriles etc., are sufficiently slow to allow for the initiation and propagation to take place, although in their presence the activity of the growing polymers is eventuelly destroyed.

Initiation of anionic polymerization of styrene, dienes and their derivatives by alkyl lithium in hydrocarbon solvents was extensively studied by Ziegler[163] and thereafter by many other workers. Since the rates of initiation are often comparable to those of propagation, both processes occur simultaneously and then, while the monomer is quantitatively polymerized, an appreciable fraction of the initiator remains unutilized in the system*. Hence, it is advantageous to use "fast" alkyl lithiums as initiators, especially when a polymer of a narrow molecular weight distribution is the desired product.

Different mechanisms govern the initiation of polymerization by alkyl lithiums depending upon whether the reaction takes place in aromatic hydrocarbons, like benzene or toluene, or whether it proceeds in aliphatic ones, like cyclo-hexane or n-hexane. The following discussion is restricted to polymerization of styrene and the dienes, and the initiation in aromatic hydrocarbons is considered first.

Polymerization of styrene initiated by n-butyl lithium in benzene was investigated by a spectrophotometric technique by Worsfold and Bywater[156]. Concentration of polystyryl anions was monitored by their absorbance at 334 nm, while the concentration of the unreacted styrene was determined by its absorbance at 291 nm. The results are shown graphically in Fig. 23. Concentration of polystyryl anions increases with time and eventually reaches its asymptotic value, being constant afterwards. This observation indicates the stability of these species. On the other hand, the concentration of styrene decays in sigmoidal fashion. This classic study unequivocally demonstrated the living character of the resulting polymers, and therefore it was justified to identify the rate of increase of the absorbance at 334 nm with the rate of initiation.

The approach of Worsfold and Bywater is superior to that of the earlier investigators who focused their attention on the overall rate of styrene consumption[157, 158] – a complex process involving simultaneous contributions from the initiation and propagation. The interest of the early workers in the maximum rate, corresponding to the slope of the sigmoidal curve at its inflection point (see Fig. 23) was unprofitable because this rate has no simple meaning. Moreover, the findings of Bywater disproved the notion of termination caused by the association of polystyryl anions with butyllithium – a hypothesis advocated by O'Driscoll and Tobolsky[157].

---

* Whatever the rate of initiation, it cannot be quantitative if the polymerization proceeds without termination. As the reaction progresses, the concentration of growing polymers increases while the concentration of the initiator decreases. Hence the former more and more efficiently compete for the monomer than the latter. However, in a fast initiation, say 99% or more of the initiator could be consumed at the early stages of polymerization and then it is appropriate to claim that the initiation quantitative (see Fig. 30)

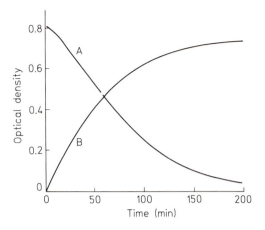

Fig. 23. Typical conversion curves recorded spectrophotometrically for initiation of styrene polymerization by n-BuLi in benzene. *Curve A:* Disappearance of styrene monitored at 291 nm. *Curve B:* Appearence of polystyryl anions monitored at 335 nm. Initial conditions [styrene] $= 1.4 \cdot 10^{-2}$ M, [n-BuLi] $= 1.1 \cdot 10^{-3}$ M

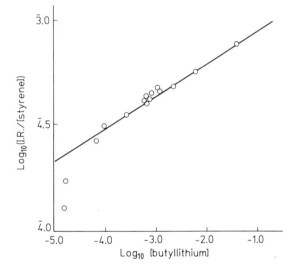

Fig. 24. Dependence of the initial rate of initiation of styrene polymerization in benzene on the concentration of n-butyl lithium. The slope of the *solid line* = 1/6

The early rate of initiation was found to be proportional to the first power of styrene concentration and to approximately ⅙ power of the concentration of n-BuLi, viz.

Rate of initiation = *const.* $[\text{BuLi}]^{1/6}[\text{styrene}]$.

This is shown in Fig. 24. In view of the hexameric nature of n-BuLi, it was plausible to rationalize this result in terms of the following mechanism:

$(n\text{-BuLi})_6 \rightleftharpoons 6\,n\text{-BuLi}$,   K ,

$n\text{-BuLi} + \text{CH}_2 : \text{CH(Ph)} \rightarrow n\text{-Bu} \cdot \text{CH}_2 \cdot \overline{\text{CH}}(\text{Ph}), \text{Li}^+$, k_i ,

that identifies the *const.* with $K^{1/6} \cdot k_i$. Results of similar studies reported in the literature make it plausible to assume the participation of monomeric n-BuLi in the initiation. For example, kinetics of addition of n-BuLi to 1,1-diphenyl-ethylene in benzene,

$n\text{-BuLi} + CH_2 : C(Ph)_2 \rightarrow n\text{-Bu} \cdot CH_2 \cdot \overline{C}(Ph)_2, Li^+$,

was found to be first order in the olefine and again ⅙ order in n-BuLi[168]. This reaction is simpler than the polymerization of styrene because 1,1-diphenyl-ethylene does not polymerize. Metalation of fluorene by n-BuLi in benzene,

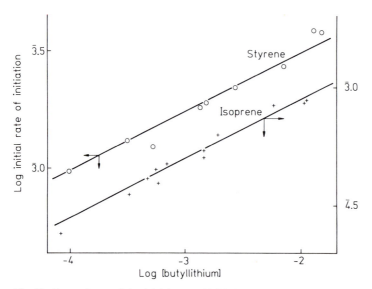

was also found to be first order in the hydrocarbon and ⅙ order in n-BuLi[170].

Kinetics of addition of the tetrameric t-BuLi to 1,1-diphenyl ethylene in benzene was investigated by Evans et al.[169]. This reaction was found to be first order in the ethylene but, significantly, ¼ order in t-butyl lithium. The ¼ order dependence of the initiation induced by the tetrameric sec-butyl lithium was observed in the polymerization of styrene or isoprene proceeding in benzene[159, 164]. This is shown in Fig. 25. Both observations lend further support to the schemes involving monomeric alkyl lithiums as the active, initiating species.

The initiation by sec-BuLi is by several powers of ten faster than by n-BuLi. Either the equilibrium concentration of the monomeric sec-BuLi is much greater than that of n-BuLi, or its reactivity is higher. The first alternative seems to be more plausible.

An interesting reagent, viz. menthyl lithium prepared by Glaze and Selman[165], is perhaps the fastest reported alkyl lithium initiator[155]. Significantly, it exists in benzene and in cyclohexane as a dimer[160]. Its extraordinary reactivity is manifested by its addition to tetraphenyl ethylene[160] – a highly shielded olefine to which no other alkyl lithium could be added.

Fig. 25. Dependence of the initial rate of initiation of styrene or isoprene polymerization on the concentration of sec.-BuLi in benzene at 30 °C. The slopes of the *solid lines* = 1/4

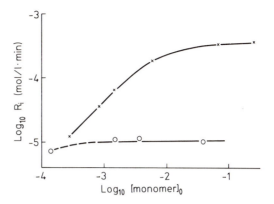

Fig. 26. Dependence of the initial rate of initiation of polymerization on the concentration of t-BuLi. Solvent: benzene. *Lower curve:* styrene, $[t\text{-BuLi}] = 4.5 \cdot 10^{-4}$ M. *Upper curve:* isoprene, $[t\text{-BuLi}] = 1.37 \cdot 10^{-3}$ M. Note the plateau of isoprene curve (-3.5) is *higher* than that of the styrene curve (-4.5) after adjustment to the concentration $[t\text{-BuLi}] = 1.37 \cdot 10^{-3}$ M. The initiation is first order in t-BuLi

The initiation of styrene polymerization by t-butyl lithium in benzene is an apparent exception. The rates were found to be proportional to the first power concentration of the lithium compound and *independent* of the monomer concentration which was varied from 0.1 to 10^{-3} M[172]. This strange behavior is illustrated by Fig. 26. The authors proposed that the dissociation of t-butyl lithium into smaller, active aggregates, dimers or monomers, is the rate determining step of this process. However, the rate of initiation of isoprene by t-butyllithium in benzene shows a more conventional behavior; it increases with increasing concentration of the monomer, although again it is proportional to *first* power concentration of t-butyl lithium.

The proposed explanation of the strange behavior of t-BuLi in the initiation of styrene polymerization in benzene is questionable because the initiation of polymerization of isoprene is *faster* than of styrene (compare the data shown in Fig. 26 after correcting them for the difference in the concentrations of t-BuLi). Therefore, an alternative explanation is advocated, namely a complexation of t-BuLi with the monomer followed by the initation,

$$(t\text{-BuLi})_4 + \text{monomer} \rightleftharpoons (t\text{-BuLi})_4(\text{monomer})$$

$$(t\text{-BuLi})_4(\text{monomer}) \rightarrow (t\text{-BuLi})_3(t\text{-Bu-monomer}^-, \text{Li}^+) \,.$$

The second step is assumed to be the rate determining one. The complexation equilibrium for styrene presumably lies far to the right. Even at concentration of styrene as low as 10^{-3} M, 90% or more of the initiator seems to be complexed with this monomer. However, the complexation is apparently less favorable for isoprene and it becomes virtually quantitative only at its concentration greater than 10^{-1} M.

The attractive scheme postulating the initiation by monomeric akyl lithium was disputed by Brown[162], who claimed that the observed rate of initiation is faster than the rate of dissociation of the hexamer into monomer. Therefore, the equilibrium

$$(n\text{-BuLi})_6 \rightleftharpoons 6\,n\text{-BuLi}$$

could not be maintained. However, this conclusion was based on the kinetic data obtained in aliphatic hydrocarbons and, as will be shown later, in those solvents the initiation is much slower than in benzene and its mechanism is drastically different from the one governing the reaction in aromatic hydrocarbons. Moreover, the erroneous value reported in the literature for the heat of dissocation of the dimeric lithium polyisoprenyl, namely $\Delta H = 37$ kcal/mol, instead of the correct value of about 12 kcal/mol[171], added to the confusion. Indeed, a reasonable estimate of the pertinent rate constants shows a self-consistency of the proposed scheme.

The character of the initiation changes as the reaction proceeds because the growing polymers combine with alkyl lithium aggregates into mixed aggregates. The latter seem to be more dissociated than the former and this facilitates the initiation. Similar behaviour is observed in the presence of Lewis bases.

An entirely different mechanism governs the alkyl lithium initiation taking place in aliphatic hydrocarbons. This may be appreciated by inspecting the schematic Fig. 27. In aromatic hydrocarbons one observes the fastest initiation at the onset of the reaction; its rate slows down as the initiation proceeds. In aliphatic hydrocarbons a long induction period is observed, the initiation starts very slowly, eventually reaches its maximum rate and thereafter slows down as the monomer is consumed. The early rate of initiation in aromatic hydrocarbons is proportional to some fractional power of initiator's concentration, the fractional power depending on the degree of alkyl lithium aggregation. The kinetic order of the early initiation in aliphatic hydrocarbons is uncertain because of its slowness. Hsieh claimed the reaction to be first order in alkyl lithium as shown by his data graphically depicted in Fig. 28[164].

Alkyl lithiums greatly differ in their abilities to initiate polymerization. This was stated earlier in the discussion of initiation taking place in an aromatic solvent. However, these differences are even greater when the reaction proceeds in aliphatic hydrocarbons as is evident from Fig. 29 showing the initiation of styrene polymerization in *cyclo*-hexane, or Fig. 30, that depicts the rate of initiation of isoprene polymerization in *n*-hexane. The crossing of the two curves revealed in Fig. 30 is instructive; it demonstrates the difficulties of comparing the initial rates when a reaction proceeds after an induction period. Therefore, it is not surprising to find a different gradation of initiating power of alkyl lithiums when gauged by different workers. For example, Worsfold and Bywater claimed the following order of decreasing initiating power for styrene polymerization in *cyclo*-hexane,

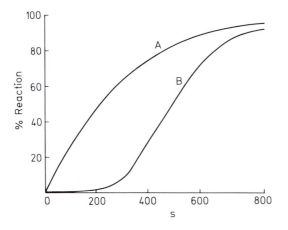

Fig. 27. Typical curves for the appearance of the absorption of polystyryl anions. *Curve A:* reaction of $1.1 \cdot 10^{-3}$ M *sec.*-BuLi with $5.3 \cdot 10^{-2}$ M styrene in benzene at 30 °C. *Curve B:* reaction of $1.3 \cdot 10^{-3}$ M *sec.*-BuLi with $8.7 \cdot 10^{-2}$ M styrene in *cyclo*-hexane at 40 °C

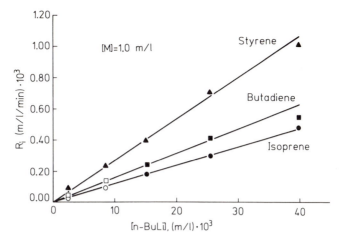

Fig. 28. Dependence of the initial rate of initiation of polymerization in *cyclo*-hexane on concentration of *n*-BuLi. Concentration of monomer 1 M. Measured by the rate of disappearance of BuLi. ▲ – Styrene, ■ – Butadiene, ● – Isoprene

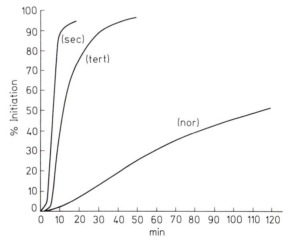

Fig. 29. Percentage of conversion of the initiator in reaction with styrene in *cyclo*-hexane at 40 °C. *Curve* (sec), [*sec.*-BuLi] = 1.2 · 10⁻³ M, [styrene] = 93 · 10⁻³ M; *Curve* (tert), [*t*-BuLi] = 1.1 · 10⁻³ M, [styrene] = 85 · 10⁻³ M; *Curve* (nor), (*n*-BuLi) = 1 · 10⁻³ M, [styrene] = 260 · 10⁻³ M. Note the three times higher concentration of styrene in reaction of *n*-BuLi

sec-BuLi > *t*-BuLi ≫ *n*-BuLi (see Fig. 29) ,

a similar order, namely

sec-BuLi > *iso*-propyl-Li > *t*-BuLi > *iso*-BuLi ≫ *n*-BuLi ,

was reported by Hsieh[164]. The order seems to depend on the nature of the monomer, e.g. in polymerization of isoprene in *n*-hexane, Bywater claimed the order

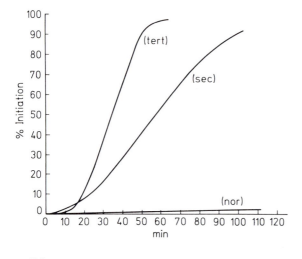

Fig. 30. Percentage of conversion of the initiator in reaction with isoprene in *n*-hexane at 30 °C. *Curve* (tert), [*t*-BuLi] = $1.0 \cdot 10^{-3}$ M, [isoprene] = $204 \cdot 10^{-3}$ M; *Curve* (sec), [*sec.*-BuLi] = $1 \cdot 10^{-3}$ M, [isoprene] = $140 \cdot 10^{-3}$ M; *Curve* (nor), [*n*-BuLi] = $0.75 \cdot 10^{-3}$ M, [isoprene] = $355 \cdot 10^{-3}$ M

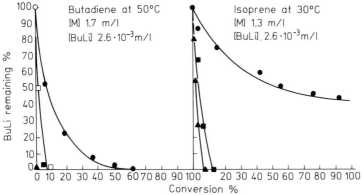

Fig. 31. Percentage of residual initiator plotted vs. percentage of monomer conversions in *cyclo-hexane*. *Left:* polymerization of 1.7 M butadiene at 50 °C, [BuLi]$_0$ = $2.6 \cdot 10^{-3}$ M. *Right:* polymerization of 1.3 M isoprene at 30 °C, [BuLi] = $2.6 \cdot 10^{-3}$ M. ▲: *sec.*-BuLi; ■: *t*-BuLi; ●: *n*-BuLi

t-BuLi > *sec*-BuLi ≫ *n*-BuLi (see Fig. 30) ,

whereas according to Hsieh the order remains unchanged. The gradation of the initiating power of alkyl lithium is even more spectacular when the fraction of the remaining initiator is compared with percent of monomer conversion. This is well illustrated in Fig. 31.

The mechanism governing the early stages of initiation in aliphatic solvents is still unknown. The concentration of the active, monomeric RLi might be exceedingly low in those solvents because the dissociation in aliphatic hydrocarbons is not facilitated by solvation. Hence, the rate of dissociation could be the rate determining step, or a slow, direct reaction of the aggregate with the monomer could start the process. In either case the rate of the early initiation would be proportional to the first power of the initiator concentration, as seems to be the case. The dependence on monomer concentration differentiates between the two alternatives discussed above. Apparently, the rate is first order in monomer[164, 175)] and this observation favors the second alternative. The bimolecular rate constant of initiation of isoprene polymerization by t-BuLi in *cyclo*hexane was reported[175] to be $8.5 \cdot 10^{-4}$ $M^{-1}s^{-1}$ at 30 °C. As the reaction proceeds mixed aggregates are formed and these are either more reactive or more readily dissociated than the homo-aggregate. This accounts for the auto-catalytic character of the initiation.

The notion of slow or low dissociation of alkyl lithiums in aliphatic hydrocarbons is supported by Brown[144], who found the intermolecular exchange between t-butyl lithium and trimethyl-silyl-methyl lithium to be extremely slow in cyclopentane. The half-life time of redistribution being 5–6 h, whereas in toluene the reaction is 10^3–10^4 times faster. The exchange in cyclopentane between the aggregated lithium polyisoprenyl and n-, sec- and t-butyl lithium was examined by Schué and Bywater[176]. The ^{7}Li-NMR technique developed by Brown was utilized. The exchange was fast with n- and sec-butyl lithium, requiring only a fraction of a second at ambient temperature. A slower exchange was observed with t-butyl lithium; nevertheless, it was completed within less than a minute*. Hence, the exchange should readily be accomplished in polymerization initiated in aliphatic hydrocarbons by n- or sec-butyl lithium. The slower exchange with t-butyl lithium should lead, under some conditions, to different kinetic behavior than that found for other systems. This indeed was observed in the polymerization of styrene and isoprene[177].

The effect of Lewis bases on the rate of initiation was noted by the early workers in the field. For example, the addition of secondary or tertiary lithium alkyls to ethylene is observed only in the presence of ethers or amines[167] The acceleration of the n-butyl lithium initiation caused by small amounts of tetrahydrofuran was reported by Dolgoplosk et al.[166]. Mixed aggregates formed with lithium alkoxides appear to be more reactive, or more readily dissociated, than the homo-aggregates. Hence, the initiation is faster in the presence of alkoxides and the induction period disappears. The accelerating effects of lithium alkoxides were reported first by Roovers and Bywater[175] and subsequently systematically investigated by Guyot and Vialle[178]. The latter workers found the n-BuOLi to be the most active in accelerating the initiation by n-butyl lithium in cyclohexane; the accelerating effect decreases when sec- or t-BuOLi is substituted for n-BuOLi. In fact, they suggest that a complex of n-BuLi and n-BuOLi provides a good initiating system for preparation of mono-dispersed and stereo-regular polyisoprene. However, they note also an inhibiting effect of the alkoxides at later stages of some reactions, e.g., n-BuOLi substantially accelerates the initiation by t-BuLi; but after a short time, it practically quenches the initiation, preventing complete utilization of the initiator. The same behavior, although less drastic, was observed with sec-BuLi. Interestingly, although lithium alkoxides accelerate the alkyl lithium initiation in aliphatic hydrocarbons[211], they markedly retard the reaction if performed in benzene solution[175].

The effect of impurities appears to be of a similar nature. For example, meticulously purified t-BuLi is a fast initiator of styrene polymerization in benzene, whereas the commercial product is a slow one. On the other hand, some impurities alleviated the induction period observed in polymerizations taking place in aliphatic hydrocarbons.

The nature of impurities is unknown. They could be volatile, e.g., in the system isoprene-n-butyl lithium in cyclohexane, the sigmoidal character of the initiation was still

---

* The rapidity of exchange of RLi's with lithium poly-isoprenyl in *cyclo*pentane, contrasting the extremely slow exchange between two kinds of RLi's in the same solvent, seems to indicate a very rapid dissociation of poly-isoprenyl aggregates (tetramers into dimers?) when compared with the dissociation of RLi's aggregates

observed when the solvent and the residual monomer were distilled from a polymerizing mixture onto fresh butyl lithium and the ensuing initiation monitored again. However, the induction period vanished on deliberate addition of oxygen at the start of the reaction.

In the presence of an excess of tetrahydrofuran, say at 0.15 M or more, the initiation of styrene polymerization by *n*-butyl lithium is virtually instantaneous[210]. The optical density at 335 nm (λ_{max} of the living polymers) reaches its maximum before any measurements can be taken. The aggregates of butyl lithium either disintegrate under the influence of the ether or become extremely reactive. The latter suggestion seems more plausible since Waack was observing still the undissociated tetrameric methyl lithium in diethyl ether.

Finally, it is proper to comment on a procedure known as the "seeding technique". The rate of initiation by alkyl lithiums in hydrocarbon solvents is often slow and comparable to the rate of propagation. Hence, new growing species continually are formed as those formed earlier increase their length. This broadens the molecular weight distribution – an undesired feature of the reaction. To avoid this difficulty, a required amount of the initiator is mixed into a small portion of the monomer to be polymerized. After a while, the length of which depends on the system, when *all* the initiator is converted into living oligomers – the seeds, the remaining monomer is added. Subsequently, a uniform polymerization ensues, yielding polymers of narrow molecular weight distribution.

It could be presumed, naively, that the initiation eventually converts all the lithium alkyls into living oligomers if one waits long enough. This is not the case in an *irreversible* propagation as could be shown by a simple kinetic argument. However, propagation involving living polymers or oligomers is always *reversible*; hence, some fraction of the introduced monomer, perhaps a minute one, remains in equilibrium with the living oligomers. As the unreacted initiator interacts with the residual monomer, depropagation of the living oligomers replenishes the losses and restores the equilibrium concentration of the monomer. Thus, the rate of the virtually quantitative conversion of the residual initiator into living oligomers is determined by the rate of depropagation.

The method utilizing monomers of low ceiling temperature e.g., α-methyl-styrene, yielding highly reactive carbanions is based on the same principle. It ensures a quantitative conversion of a slow initiator into a highly reactive species that, in turn, rapidly initiates polymerization of the required monomer.

Lithium alkyls initiate polymerization of polar monomers like methyl-methacrylate, vinyl pyridine, acrylo-nitrile, etc. However, these reactions are more complex. The desired addition to the C=C bond is accompanied by other processes, e.g., attack on the –COOEt group with the formation of ketones and lithium methoxide, or in vinyl pyridine polymerization by the metalation of the pyridine moiety.

Although a conventional initiation of anionic polymerization by alkyl lithiums involves their addition to a C=C bond, an exceptional mode of initiation associated with ring-opening polymerization was reported by Morton and Kammereck[186]. Ethyl lithium seems to initiate polymerization of propylene sulphide, but the reaction involves first a desulphuration of the monomer, leading to the formation of lithium mercaptide:

$$
\underset{\displaystyle \overset{\textstyle CH_3}{\underset{\textstyle S}{CH-CH_2}}}{} + \ EtLi \ \longrightarrow \ CH_3CH{:}CH_2 \ + \ EtSLi
$$

which is the actual initiator:

$$\underset{\underset{S}{\diagdown\diagup}}{\overset{CH_3}{\underset{|}{CH}}\text{—}CH_2} + EtSLi \longrightarrow EtS \cdot CH_2\text{–}\overset{CH_3}{\underset{|}{CH}} \cdot SLi, \text{ etc.}$$

On the other hand, lithium mercaptides do not initiate polymerization of the analogous 4-member cyclic sulphides and their reaction with ethyl lithium proceeds in a conventional way:

$$\overset{CH_3}{\underset{\underset{CH_2\text{—}CH_2}{|}}{\underset{|}{CH}}\text{—}S} + EtLi \longrightarrow EtS \cdot \overset{CH_3}{\underset{|}{CH}} \cdot CH_2CH_2Li, \text{ etc.}$$

Neither 5- or 7-membered cyclic sulphides could be polymerized.

The initiation of propylene sulphide polymerization by α-sulphonyl type carbanions, $-SO_2, CH_2Li$, proceeds in a conventional way[187].

III.9. Initiation of Anionic Polymerization by Lewis Bases

Anionic polymerization initiated by uncharged nucleophiles has been observed in several systems. The first example of such a reaction probably was furnished by the work of Horner et al.[236], who initiated polymerization of nitroethylene and of acrylonitrile by trialkyl phosphines. The following mechanism was postulated:

$$R_3P + CH_2:CHNO_2 \rightarrow R_3\overset{+}{P} \cdot CH_2\overline{C}HNO_2$$

followed by the propagation taking place on the $-\overline{C}HNO_2$ anion. The resulting polymer was visualized as a macro-zwitter-ion.

The same mechanism was postulated by Katchalsky[234, 235] for polymerization of nitroethylene initiated by pyridine or its derivatives.

Only oligomers of DP $\sim$ 10–15 were formed at ambient temperature, and since their molecular weight was only slightly affected by the monomer concentration, a chain transfer to monomer was proposed. Kinetics of the polymerization were studied by following its progress through temperature rise in an adiabatic enclosure. The overall reaction was found to be first order in pyridine but second order in the monomer, implying a slow bimolecular initiation. By combining the kinetic data with those obtained from the dependence of the degree of polymerization on the monomer concentration, the rate of initiation was evaluated,

$$Py + CH_2:CHNO_2 \rightarrow \overset{+}{P}y \cdot CH_2 \cdot \overline{C}HNO_2 , \quad k_i \sim 4 \cdot 10^{-2} \text{ M}^{-1}\text{s}^{-1}$$

at 20 °C in methyl-ethyl-ketone.

The polymerization performed in dimethyl-formamide or in tetrahydrofuran was investigated over a wide temperature range. At low temperatures, viz. -104 to $-75\,°C$, the molecular weight of the polymer increased with conversion, implying living character of the polymerization. In fact, a polymer of molecular weight as high as 60,000 was obtained at $-104\,°C$. Unfortunately, it was not established whether the pyridine moiety was still present in those polymers; hence, a plausible initiation scheme involving proton-transfer from the monomer to the primary zwitter-ion could not be excluded:

$$\overset{+}{P}yCH_2\overset{-}{C}HNO_2 + CH_2:CHNO_2 \rightarrow \overset{+}{P}yCH_2CH_2NO_2 + CH_2:\overset{-}{C}NO_2 \,,$$

$$CH_2:\overset{-}{C}NO_2 + CH_2:CHNO_2 \rightarrow CH_2:C(NO_2)\cdot CH_2\cdot \overset{-}{C}HNO_2 \text{ etc.}$$

The anionic polymerization of acrylonitrile initiated by triphenylphosphine was re-investigated by Jaacks et al.[238]. The absence of the phosphine in the polymer and the isolation of 99% of the initiator in the form of the phosphonium salt, $Ph_3\overset{+}{P}CH_2CH_2CN$, demonstrates that macro-zwitter-ions are not formed by this initiation, a conclusion contrary to the claims of other workers, e.g. Enikolopyan et al.[244]. The previously discussed proton-transfer was established therefore,

$$Ph_3\overset{+}{P}CH_2\overset{-}{C}H(CN) + CH_2:CH(CN) \rightarrow Ph_3\overset{+}{P}CH_2CH_2CN + CH_2:\overset{-}{C}CN \,,$$

and its rate seems to be much faster than of the competing monomer addition,

$$Ph_3\overset{+}{P}\cdot CH_2\cdot \overset{-}{C}H(CN)\cdot CH_2\cdot \overset{-}{C}H(CN) \rightarrow Ph_3\overset{+}{P}\cdot CH_2\cdot CH(CN)\cdot CH_2\cdot \overset{-}{C}H(CN)\cdot \,.$$

However, the addition of the monomer to the $CH_2:\overset{-}{C}(CN)$ or $\sim CH_2\overset{-}{C}H(CN)$ anions is faster than the analogous proton-transfer. This is not surprising. The close vicinity of the positive charge to the carbanion of the primary zwitter-ion makes it a much weaker nucleophile than the macro-carbanion, hence the monomer addition is impeded. On the other hand, in many proton-transfer reactions tight ion-pairs are more reactive than the loose ones or the free carbanions. Since the primary zwitter-ion could be treated as a very tight ion-pair, its reactivity in proton-transfer might be very high.

Strangely enough, the initiation of acrylonitrile polymerization by triethylphosphite, $(EtO)_3P$, is claimed to produce polymers endowed with the phosphite end-groups[237]. These polymerizations were carried out at high concentrations of the phosphite, e.g., in some runs the ratio of phosphite/acrylonitrile was $1:1$. Nevertheless, high-molecular weight polymers, e.g. of DP as high as 280, were obtained at $-10\,°C$. Their analysis showed the presence of 1 phosphorus atom per 1 polymer molecule. However, this ratio decreases at higher temperatures, e.g. to 0.1 at $50\,°C$.

Spectroscopic observations revealed a reversible formation of charge-transfer complex between the phosphite and acrylonitrile,

$$(EtO)_3P + CH_2:CH(CN) \rightleftharpoons (EtO)_3P\cdots CH_2:CH(CN) \quad \text{c.t. complex.}$$

A similar complex was observed on mixing triphenylphosphite with acrylonitrile although, in spite of its formation, polymerization did not ensue* [239]. Hence, the following mechanism was proposed for the initiation by the trietylphosphite:

(a) $(EtO)_3P + CH_2 : CH(CN) \rightleftharpoons$ charge-transfer complex ,   K

(b) charge-transfer complex $\rightarrow (EtO)_3\overset{+}{P} \cdot CH_2 \cdot \overline{C}H(CN)$, k'

(c) $(EtO)_3\overset{+}{P}CH_2\overline{C}H(CN) + CH_2 : CH(CN) \rightarrow (EtO)_3\overset{+}{P}CH_2 \cdot CH(CN) \cdot CH_2 \cdot \overline{C}H(CN)$, k'' .

Kinetics of the overall reaction proceeding in ethanol was investigated[239, 249]. A pseudo-first order termination was assumed to take place in this solvent, namely,

$EtOH + (EtO)_3\overset{+}{P} \cdot CH_2CH(CN) \wedge\wedge\wedge CH_2 \cdot \overline{C}H(CN)$

$\rightarrow EtOEt + (EtO)_2PO \cdot CH_2CH(CN) \wedge\wedge\wedge CH_2 \cdot CH_2(CN)$, k_t ,

and indeed formation of diethyl ether was noted. At lower temperatures reaction (b) was assumed to be the rate determining, i.e., the primary zwitter-ions are consumed by reaction (c) as soon as formed. Hence, the rate of polymerization, R_p, in the stationary state is given by

$$R_p = (k'k_pK/k_t)[(EtO)_3P][CH_2 : CH(CN)]^2 ,$$

in agreement with the observations. At higher temperatures a third order dependence on monomer concentration was claimed. However, this might be an artifact resulting from a substantial loss of low-molecular weight polymers at lower concentrations of the monomer.

The previously discussed proton-transfer from the monomer to the primary zwitter-ion, e.g., $R_3\overset{+}{P} \cdot CH_2\overline{C}CH(CN)$, involves a proton on that carbon atom of the monomer that bears the substituent. Such protons are unavailable in vinylidene monomers; therefore, macro-zwitter-ions could be formed in such systems. Jaacks and Franzmann[241] showed this to be the case in the polymerization of methylene-malonic ester initiated by triphenyl phosphine. Relatively highmolecular weight oligomers, $\sim 5,000$, were obtained by terminating the polymerization with $K_2S_2O_7$. This introduces a stable anionic end-group, $-CH_2C(COOR)_2SO_3^-$, and preserves the zwitter-ionic structure of the polymer. After fractionation and rigorous removal of low-molecular weight material, the absorption spectrum of the oligomers showed a band at $\lambda_{max} = 268.6$ nm, identical with that observed in the spectrum of a model compound,

$Ph_3\overset{+}{P} \cdot CH_2CH(COOR)_2, Br^-$.

Hence, the presence of the phosphonium end-groups in this polymer was documented. Their proportion is high in the purified oligomers, about 60%.

---

* Polymerization ensues on UV irradiation of that complex. Its mechanism is not well established. The authors proposed that the photolysis of that complex yields free radicals and the latter initiate radical polymerization of acrylonitrile

The large proportions of the non-utilized phosphine and of the primary zwitter-ions in the low-molecular weight residue prove two points:

(a) The addition of the phosphine to methylene malonic ester is slow;

(b) The conversion of the primary, monomeric zwitter-ion into a dimeric one is also slow, much slower than the subsequent propagation.

This is plausible. The separation of charges is more difficult the closer they are to each other. The oligomeric chains, up to DP = 8, are too stiff to allow for cyclization; hence, Jaacks' work demonstrates the feasibility of charge separation, even in a moderately polar solvent. Such a separation could be facilitated by a mutual association of the propagating zwitter-ions.

Ring-opening polymerization of β-lactones, such as β-propiolactone or pivalactone, by tertiary amines was studied by several research groups. The first initiating step leads to the formation of betaines[245], e.g.,

$$R_3N + \underset{\underset{CH_2-O}{|}}{CH_2-CO} \longrightarrow R_3\overset{+}{N}CH_2CH_2COO^-$$

and these could propagate the polymerization of lactones by a nucleophilic attack of their carboxylate ions on the uncoming monomer. The resulting polymer should have the zwitter-ionic character.

The feasibility of the zwitter-ionic mechanism was questioned by some investigators who visualized alternative modes of initiation yielding free and unbound cations. The previously discussed proton-transfer from a monomer is a possibility; Hofmann degradation of the betain provides another alternative[248]:

$$R_3\overset{+}{N} \cdot CH_2CH_2COO^- \rightarrow R_3\overset{+}{N}H + CH_2:CH \cdot COO^- .$$

In either case the acrylate anion would be the ultimate initiator, yielding a conventional, non-zwitter-ionic polymer. It was necessary, therefore, to provide evidence for the presence of the ammonium end-groups in the polymers, and this was done by Jaacks and Mathes[240] through NMR and chemical analysis of fractionated and rigorously purified oligomers. However, on prolonged storage of the polymer, the ammonium end-groups are readily lost, presumably due to Hofmann degradation, especially at higher temperatures. The failure of Yamashida et al.[248] in detecting the ammonium end-groups in their polyesters is not surprising since the polymers were heated up to 90 °C.

Further evidence of direct initiation by betaines was provided by the investigations of β-propiolactone polymerization initiated by the $Me_3N^+CH_2COO^-$ betaine[242]. The absence of β protons in that betaine prevents its Hofmann degradation and facilitates studies of the macro-zwitter-ions. The polymerization was carried out in ethanol. By choosing a high ratio of the betaine to the lactone, zwitter-ions of low degree of polymerization (~11) were obtained. The nitrogen content of the purified oligomers and their electrophoresis after estrification of the terminal carboxylate groups left no doubt about their structure.

Kinetics of β-propiolactone polymerization initiated by the $Me_3\overset{+}{N} \cdot CH_2COO^-$ betaine in absolute alcohol was followed by the IR technique. The betaine, lactone and the polymer are all soluble in alcohol and hence the reaction is homogeneous. Unfortunately, chain transfer to the solvent complicates its course; nevertheless, it was possible to show that the rate constant of initiation,

$$\overset{+}{Me_3NCH_2CH_2COO^-} + \underset{\underset{CH_2-O}{|}}{\overset{\overset{CH_2-CO}{|}}{}} \longrightarrow \overset{+}{Me_3NCH_2CH_2COOCH_2CH_2COO^-}$$

is about four times slower than the rate constant of the subsequent propagation.

Zwitter-ionic character of anionic polymerization of formaldehyde is claimed by Buehlke et al.[243] but details of that study are not available.

Formation of macro-zwitter-ions was demonstrated by UV analysis in the polymerization of cyano-acrylate esters initiated by trietyl- or triphenyl-phosphine, or alternatively by pyridine or tertiary amines[250]. The cyanoacrylate esters, like vinylidene cyanide, are extremely reactive monomers; their rapid polymerization is easily initiated by traces of weak bases, e.g. water. The electron-withdrawing substituents greatly stabilize the growing carbanions, making them resistent to termination. The initiation by the phosphines is irreversible and rapid, even when compared with the fast propagation. The subsequent monomer addition to the primary zwitter-ions is also fast. Hence, the utilization of the initiator is quantitative and the overall polymerization is first order in the phosphine and in the monomer. At lower temperatures the ensuing polymerization exhibits all the features of an ideal living polymerization.

Initiation by pyridine is much slower and its consumption is very low. The rate of the ensuing polymerization increases with decreasing temperature implying a reversible, exothermic addition of pyridine to the monomer. Plots of $[monomer]_t/[monomer]_0$ versus time are convex; the slopes at $t = 0$ increase with increasing initial monomer concentration – evidence of a slow second step of initiation:

$$\overset{+}{Py}CH_2\overset{-}{C}(CN)(COOR) + CH_2 : C(CN)(COOR) \rightarrow \overset{+}{Py}CH_2C(CN)(COOR)CH_2\overset{-}{C}(CN)(COOR) .$$

Consequently, the overall initiation is second order in monomer.

The reversibility of the addition of tertiary amine is even more pronounced. The initiation is very slow and insignificant towards the end of polymerization when the concentration of monomer is greatly reduced. Therefore, the conversion curves at that stage of the process show an internal first order character in monomer but second order in respect to the initial monomer concentration.

Most interesting systems referred to as "no catalyst polymerization" were discovered and comprehensively investigated by Saegusa[291]. The reaction results from mixing of two monomers of greatly different polarity, the electron donating one, M_E, and the electron accepting M_N. Their interaction leads to formation of zwitter-ions, $^+M_E \cdot M_N^-$, which in turn rapidly polymerize, yielding alternating co-polymers. Although some macro-cycles were found in the products, most of the polymerized material seems to be linear. The question of how the cyclization is prevented in the early stages of polymerization remains to be answered. This interesting field is reviewed by Saegusa[296, 503].

III.10. Charge-Transfer Initiators and Simultaneous Anionic and Cationic Homo-Polymerizations

In a paper devoted to the investigation of radical co-polymerization of vinylidene cyanide with a variety of monomers, an unusual autocatalytic behavior was reported for the pair: vinylidene cyanide – vinyl ether[251]. The authors suggested a possible mutual co-initiation of anionic polymerization of vinylidene cyanide, inducing a cationic polymerization mode of vinyl ether.

The idea of two simultaneous homo-polymerizations was explored by Szwarc and his co-workers[252]. Pac and Plesch[253] reported initiation of cationic polymerization of vinyl carbazole (VC) by tetranitromethane in nitrobenzene. They postulated the formation of a charge-transfer complex,

$$C(NO_2)_4 \cdot VC \; ,$$

which was assumed to dissociate into $C(NO_2)_4^- \cdot$ radical anion and VC^+ radical cation, the latter initiating the observed polymerization. Such a mechanism suggests a possible simultaneous anionic and cationic polymerization of two judiciously chosen monomers. Attempts by Szwarc et al.[254] to induce in that system anionic polymerization of nitroethylene were in vain. On the other hand, they observed formation of a charge-transfer complex on mixing tetranitromethane with 1,1-diphenyl ethylene followed by its slow decomposition into a $C(NO_2)_3^-$ anion and a nitrated 1,1-diphenyl ethylene cation

$$NO_2 \cdot CH_2 \cdot \overset{+}{C}(Ph)_2 \; .$$

The anion deprotonates the cation; hence, nitroform, $HC(NO_2)_3$, and the nitrated 1,1-diphenyl ethylene, $NO_2CH : C(Ph)_2$, are the ultimate products of this process. It was proposed, therefore, that the transfer of NO_2^- cation, and not electron transfer, yields nitroform and nitrated vinyl carbazole,

and hence the polymerization of vinyl carbazole observed by Plesch ensues from proton transfer with nitroform – a relatively strong acid – acting as the proton donor.

Reinvestigation of Plesch's work confirmed their kinetic conclusions, viz. the first order kinetics in monomer, and led to a similar rate constant. However, addition of oxetane to the vinyl-carbazole-tetranitromethane system greatly retarded the vinyl-carbazole polymerization, e.g. by a factor of 30, but neither the first order character of this reaction nor the molecular weight of the resulting polymer was affected provided that the concentration of oxetane was sufficiently large. At the same time, cationic polymerization of the added oxetane was observed.

Polymerization of oxetane in nitro-benzene is initiated by nitroform but not by tetranitromethane although the latter initiates the polymerization on addition of carbazole or 1,1-diphenyl ethylene. It follows that oxetane and vinyl carbazole compete for the nitroform, the former more strongly than the latter. Since in the system vinyl carbazole – tetranitromethane – oxetane two homopolymers are formed, and not a block or co-polymer, and because the presence of oxetane does not affect the molecular weight of the poly-vinyl carbazole, one concludes that the growing poly-vinyl carbazole cation does not initiate oxetane polymerization. Apparently, this cation is too inert to react with oxetane but it adds vinyl carbazole due to its high polarizibility that facilitates the reaction. On the other hand, the cyclic oxonium ion of the growing poly-oxetane is not a sufficiently strong electrophile to attack vinyl carbazole. These assumptions could account for the simultaneous two homo-propagation.

Simultaneous co-initiation of anionic and cationic polymerization of vinylidene cyanide and of various cyclic ethers, e.g. oxetane or tetrahydrofuran, was reported by Stille et al.[255]. On mixing a solution of vinylidine cyanide with an appropriate cyclic ether, a spontaneous polymerization of both monomers ensues yielding two homo-polymers and not a block or a co-polymer. For a constant concentration of vinylidene cyanide its rate of polymerization increases, while the molecular weight of its polymer decreases, with increasing concentration of the ether. A corresponding relation was found in the reverse case. Phosphorous pentoxide – a typical inhibitor of anionic polymerization – inhibits the formation of poly-vinylidene cyanide, confirming its anionic character. In analogy, addition of pyridine – a typical inhibitor of cationic polymerization – prevents the polymerization of the cyclic ether, proving its cationic character.

Vinylidene cyanide is much more rapidly polymerized than the cyclic ether, as shown in Fig. 32. Furthermore, it precipitates as it is formed, burrying its anionically growing end-group in the resulting gel and making it inaccessible to the cationically growing poly-ether. Hence, these two polymerizations proceed in different phases and, to a great extent, at different times.

The following mechanism of initiation was proposed:

$$\square\hspace{-0.5em}O + CH_2{:}C(CN)_2 \longrightarrow \square\hspace{-0.5em}\overset{+}{O}-CH_2\cdot\bar{C}(CN)_2$$

The anionic end of the resulting zwitter-ion grows rapidly while the cationic end extracts hydride from THF, yielding

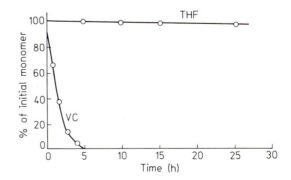

Fig. 32. Rates of disappearance of vinylidene cyanide (VC) and tetrahydrofuran (THF) in spontaneous polymerization of their 1 : 1 mixture of 20 °C

The deprotonation yields dihydrofuran and a secondary oxonium ion[256] – the actual initiator of the ether polymerization,

This speculative mechanism needs verification, and the problem of simultaneous co-initiation deserves further investigation.

III.11. Electro-Chemical Initiation

Electro-chemically initiated anionic polymerization was investigated by Funt and his co-workers[206, 293]. For example, living polystyrene was generated by electrolysis of styrene in tetrahydrofuran with $NaBPh_4$ or $NaAlEt_4$ as the supporting electrolyte. It is necessary to use in this process a divided electrolytic cell because the products of anodic oxidation, presumably the radicals $BPh_4^{.}$, or $AlEt_4^{.}$, act as powerful terminating agents. Electron-transfer from the cathode to the monomer is assumed to be the initiating step.

A similar procedure led to polymerization of α-methyl styrene[207], while Bhadani and Parravano[208] reported a rapid and quantitative polymerization of 4-vinyl-pyridine. The latter reaction was initiated by cathodic reduction of the monomer in hexamethyl-phosphorictriamide with $NaBPh_4$ as the supporting electrolyte.

It has been questioned whether the initiation involves a direct electron-transfer from a cathode to a monomer or whether the electrolysis leads to the deposition of sodium metal on the cathode and the deposited metal is then the initiating species. This question was answered by using quartenary ammonium salts as the supporting electrolytes[209]. Since this change of the supporting electrolyte did not affect the results, a direct electron-transfer to monomer had to be responsible for the initiation. In fact, Parravano pointed out that the applied potential is not sufficiently high to allow the deposition of metallic sodium. Application of electro-chemical initiation in a quantitative study of propagation is reviewed on p. 132.

An interesting electro-reduction leading to radical polymerization was reported recently by Delamar[420]. Perchlorate anions are reduced on cathode to the respective radicals which initiate polymerization. Electro-chemical initiation is interesting; it provides a way for continuous change of the redox potential of the initiating agent. However, there are many restrictions which limit its usefulness. The solvents must be capable of dissolving the monomer, polymer, and the supporting electrolyte, and to allow for a measurable dissociation of the salt into free ions. Moreover, neither the solvent nor the supporting electrolyte should be attacked by the resulting living polymers, and the cations of the supporting electrolyte must have a higher reduction potential than the monomer. In view of all these restrictions, the electro-chemical initiation is presently confined to the realm of laboratory research.

A comprehensive review of electro-chemical initiation, including the anionic mode of propagation, was published by Breitenbach et al.[260].

III.12. Initiation of Polymerization by Ionizing Radiation

Irradiation of any material by γ-rays, short X-rays, fast electrons, etc. results in ionization of composing it molecules. The primary event consists of electron ejection which es take place, e.g. the neighboring but oppositely charged ionic species collide, neutralize their charges in reactions that yield excited molecules, fragmentation products ralize their charges in reactions that yield excited molecules, fragmentation products including free radicals and atoms, etc. In fact, most of the primary ionic species rapidly loose their charges through such a neutralization, and only a small fraction escapes their mutual attraction, diffuses into the bulk of the material and retains their charge for about a few microseconds, provided that the intensity of the radiation is not too high.

Ionizing irradiation of monomers, in bulk or in solution, leads to their polymerization. The formed free radicals, as well as the ionic species that escaped their mutual attraction, may act as the initiators. High molecular weight polymers are formed whenever the lifetime of growing species is sufficiently long and their rate of growth sufficiently fast, i.e., $k_p[M] \cdot \tau > \sim 100$, k_p denoting the relevant propagation constant and τ the lifetime of a growing species.

As pointed out long ago[462a], the ionic growth is due to the growing free ions. Encounters of two oppositely charged species do not yield ion-pairs but neutralize their charges and terminate the ionic growth. In other words, a negatively charged species encountered by a cationically growing polymer is not likely to be a "good" counter-ion, and *vice-versa*.

Propagation constants of cationically growing centers are large, e.g. k_p of a cationically growing free polystyryl ion is $\sim 10^8$ $M^{-1}s^{-1}$[464]. In bulk styrene, i.e., for $[M] \sim 10$ M, and for $\tau \sim 1$ μs, about 1,000 monomeric molecules are added to each growing center prior to its termination. The k_p of growing free polystyryl anions is substantially smaller, probably not greater than 10^5 $M^{-1}s^{-1}$. Consequently, the anionic growth, that simultaneously proceeds with the cationic propagation, contributes less than 1% to the polymer production, i.e. it remains virtually undetected.

The lifetime of ionically growing species depends on the intensity of radiation, being inversely proportional to its square-root, provided that encounters between the oppositely charged species are the only mode of termination. However, termination of ionic growth could also result from a reaction of a growing carbenium ion with some neutral species that converts a reactive cation into an unreactive one. For example, moisture present in a monomer converts the carbenium ion into an unreactive H_3O^+ ion, thus terminating the cationic growth.

$$\text{Ɱ } CH_2 \cdot \overset{+}{C}H(Ph) + H_2O \rightarrow \text{Ɱ } CH_2 \cdot CH(Ph)\overset{+}{O}H_2 \tag{a}$$

followed by

$$\text{m} CH_2 \cdot CH(Ph)\overset{+}{O}H_2 \rightarrow \text{m} CH:CH(Ph) + H_3O^+ \tag{b}$$

or

$$\text{m} CH_2 \cdot CH(Ph)\overset{+}{O}H_2 + H_2O \rightarrow \text{m} CH_2 \cdot CH(Ph)OH + H_3O^+ . \tag{c}$$

The last reaction that consumes water is important at its relatively high concentrations, and it could be visualized as a reaction of carbenium ion with $(H_2O)_2$. Since H_3O^+ regenerates water on protonating the negatively charged species, the mere irradiation cannot remove the last traces of moisture. However, in a meticulously dry monomer, the termination by water is too slow to be of significance, and the termination arising from encounters with the negatively charged ions is the only one of importance. Under such conditions the yield of cationically formed polymers is greatly increased, and this mode of polymerization becomes more important than the radical one.

Cationic polymerization also prevails over the radical one when a monomer is incapable of being polymerized by a radical mechanism. Isobutene exemplifies such a monomer. By the same token, anionic polymerization becomes the dominant one when a monomer is not polymerized by a cationic mechanism but readily polymerizes anionically. An example is provided by anionic polymerization of nitroethylene initiated by ionizing radiation[466].

IV. Propagation of Anionic Polymerization

IV.1. General Considerations

In polymerization processes yielding high-molecular weight polymers, the rate of mono-mer consumption is virtually equal to the rate of propagation,

$$- d[M]/dt = k_p[P^*][M] \quad \text{or} \quad - d\ell n[M]/dt = k_p[P^*] \; ,$$

where M denotes a monomer, P^* a growing polymer, and k_p the observed propagation constant. The concentration of growing polymers is stationary during most, but not all, polymerizations. In those processes that involve continuous initiation and termination, the stationary condition requires balancing of the rates of those two reactions. In termi-nationless systems, $[P^*]$ remains constant after cessation of initiation. However, $[P^*]$ need not be equal to *initial* concentration of the initiator since a fraction, often a large one, of the formed polymeric molecules may be in a dormant state.

A variable $[P^*]$ is observed in living polymer systems when the rate of initiation is comparable to, or slower than the rate of propagation. Plots of monomer concentration *vs.* time are then sigmoidal; the rate of monomer consumption increases as more growing polymers are formed, but eventually it decreases in the wake of depletion of the mono-mer needed for the propagation. The maximum rate observed at the time coinciding with the inflection point of the conversion curve has no simple meaning although early inves-tigators[157, 158] stressed its significance. The polymerizations involving extremely slow, and especially autocatalytic initiation exhibit induction periods. The consumption of monomer is too low in the early stages of such processes to be observed.

In ionic polymerizations the non-terminated end-groups may exist in a variety of *interchangeable* states such as of free ions, various ion-pairs, triple ions, quadrupoles, etc. Each polymeric end-group exists in each accessible state for some time; and in that period it contributes to propagation as required by its momentary character. Hence, the observed propagation constant k_p is given by

$$k_p = \sum k_{pi} f_i \; ,$$

where k_{pi} is the propagation constant of polymeric end-groups in state i, f_i denotes the mole fraction of polymers present in that state, and the summation is carried over all the accessible states. In most of the conventional polymerizations, the mole fractions, f_i's, are determined by the equilibria established between the inter-convertable states, viz. by the reactions

$$P_i^* \; \underset{k_{ji}}{\overset{k_{ij}}{\rightleftharpoons}} \; P_j^* \;.$$

Propagation does not perturb such equilibria, provided that the addition of monomer to an n-meric species i again yields a species i but longer by one unit, i.e.

$$P_i^* + M \; \underset{k_{di}}{\overset{k_{pi}}{\rightleftharpoons}} \; P_i^* \;.$$

n-mer $\qquad\qquad\quad n+1$-mer

In a contrary case, when propagation could convert a species i into a species j and depropagation could yield a species i from a species j, viz.

$$P_{i,n}^* + M \; \underset{k_{dji}}{\overset{k_{pij}}{\rightleftharpoons}} \; P_{j,n+1}^* \;,$$

n-meric $\qquad\qquad\quad n+1$-meric

the respective mole fractions are determined by the stationary state and then their values are affected by the concentration of monomer.

The inter-conversion of the polymeric species P_i^* into P_j^* could be spontaneous or could involve an external agent. For example, free ions are converted into ion-pairs by being recombined with appropriate counter-ions, viz.

$$P^- + Cat^+ \rightleftharpoons P^-, Cat^+ \;.$$

An ion-pair not associated with a solvating agent L could be converted into an ion-pair associated with L, like

$$P^-, Cat^+ + L \rightleftharpoons P^-, Cat^+, L \;.$$

Since the concentrations of the external agents could be varied at will, independently of the concentrations of P_i^*'s, the concentrations and lifetimes of the pertinent polymeric species could be modified according to needs of an investigator.

Cations and solvents often exert a decisive influence on the course of anionic propagation. For example, it was pointed out earlier that methyl-vinyl-sulphone is not polymerized when metallic potassium is used for the initiation. Under those conditions a radical-anion is formed which abstracts a proton from another molecule of the sulphone, ultimately yielding a salt

$$CH_2 : CH \cdot SO_2 \cdot CH_2^-, K^+ \;.$$

This salt was claimed not to be sufficiently basic to propagate the polymerization. However, this sulphone polymerizes, yielding a rather high-molecular weight product, when the reaction is initiated by n-BuLi in THF or DME[258, 259]. It was proposed that the resulting n-Bu $\cdot$ CH$_2$ $\cdot$ $\overline{\text{C}}$HSO$_2$CH$_3$, Li$^+$ pair is capable of propagating the polymerization, in contrast to the CH$_2$: CH $\cdot$ SO$_2$ $\cdot$ $\overline{\text{C}}$H$_2$, K$^+$. However, this writer suspects that the different behavior is caused by the change of cation. Numerous examples revealing the sometimes dramatic effects caused by variations of solvent and/or counter-ion will be discussed later.

Finally, growing ion-pairs often aggregate and form inert associates, e.g. (P$^-$, Li$^+$)$_2$. It seems that the unassociated pairs, being in equilibrium with the associates, are the growing species.

IV.2. Molecular Weight Distribution

The participation of inter-convertable species in ionic polymerizations, each characterized by its own propagation constant, affects molecular weight distribution of the resulting polymer. For example, consider a polymerizing system involving a rapidly growing species P$^*_A$ and a slowly growing species P$^*_B$. For the inter-conversion,

$$P^*_A \rightleftharpoons P^*_B ,$$

which is very slow compared with the time of propagation, the resulting product aquires a bimodal distribution, i.e. forms a mixture of oligomers and high-molecular weight polymers. However, a uniform polymer is formed when the inter-conversion is fast.

Mathematical treatments of molecular weight distribution of polymers produced in such systems were outlined by Fox and Coleman[215], Figini[216, 230], and by Szwarc and Hermans[217]. The simplest scheme is described by the following set of reactions:

$$P^*_{A,n} + M \rightarrow P^*_{A,n+1} , \; k_{pA} ,$$

$$P^*_{B,n} + M \rightarrow P^*_{B,n+1} , \; k_{pB} ,$$

$$P^*_{A,n} \underset{k_b}{\overset{k_f}{\rightleftharpoons}} P^*_{B,n} , \qquad K = k_f/k_b .$$

Figini's treatment gives the weight average degree of polymerization of a high-molecular weight product, namely

$$\overline{DP}_w = 1 + \overline{DP}_n + (\eta + 1)(1 - 2/\varepsilon)\overline{DP}_n$$

where

$$\varepsilon = 1 + (k_f + k_b)^2/(k_{pA}k_b + k_{pB}k_f)([P^*_A] + [P^*_B]) ,$$

$$\eta = \{(\varrho^2 + K)(1 + K)/(\varrho + K)^2 + 1 - \varepsilon\}/(\varepsilon - 2) ,$$

and

$$\varrho = k_{pA}/k_{pB} .$$

As expected, for $k_{pA} = k_{pB}$, i.e. $\varrho = 1$, $\overline{DP}_w = \overline{DP}_n + 1$, a relation characteristic of *Poisson distribution*. Indeed, under this condition the polymerizing system behaves as if it were composed of only a single species and, since termination and chain-transfer are absent, Poisson distribution is obtained – a point stressed by Flory long ago[220].

For Poisson distribution, one gets:

$$\text{mole fraction of } n\text{-mers} = e^{-\gamma}\gamma^{n-1}/(n-1)\,!$$

and

$$\text{weight fraction of } n\text{-mers} = [\gamma/(\gamma+1)]ne^{-\gamma}\gamma^{n-2}/(n-1)\,!$$

where $\gamma = \overline{DP}_n - 1$.

Hence, $\overline{DP}_w = \gamma + 1 + \gamma/(\gamma+1)$ and $\overline{DP}_w/\overline{DP}_n = 1 + \gamma/(\gamma+1)^2$.

For $\gamma \gg 1$, the ratio $\overline{DP}_w/\overline{DP}_n \sim 1.0$.

The degree of uniformity attained in polymers characterized by Poisson molecular weight distribution increases with increasing degree of polymerization. For example, for $\overline{DP}_n = 100$ about 70% of polymers have molecular weights within 10% of their average value. Their proportion increases to 96% for $\overline{DP}_n = 500$. The corresponding ratios $\overline{DP}_w/\overline{DP}_n$ are 1.01 and 1.002, respectively. For the sake of illustration, the dependence of weight fraction of n-mers on their degree of polymerization, n, is shown in Fig. 33 for different $\overline{DP}_n$'s. The following experimental conditions have to be met to obtain polymers characterized by Poisson molecular weight distribution:
(1) The polymerization has to be irreversible.
(2) Any termination and chain-transfer have to be rigorously excluded.
(3) The initiation has to be virtually instantaneous, or low-molecular weight oligomers have to be used as initiators.
(4) Identical conditions should be maintained throughout the whole reactor, i.e. the concentration and temperature should be uniform. This requires an efficient stirring of the polymerizing solution, more easily achieved in a slow polymerization than in a fast one

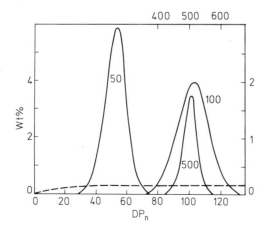

Fig. 33. Poisson distribution of molecular weights. Plot of weight fraction vs. degree of polymerization. The average degrees of polymerization of the three curves shown in the figure are 50, 100, 500

As pointed out by Szwarc and his co-workers[221, 222], anionic polymerization yielding living polymers is capable of fulfilling all these conditions. Indeed, after overcoming various technical difficulties, many workers prepared by this technique high-molecular weight polymers characterized by Poisson molecular weight distribution.

Whenever a polymerization involves chain-transfer or termination, a product having the most probable molecular weight distribution is obtained, provided that p, the ratio of probabilities of growth and transfer or termination remains constant throughout the duration of the process. Under such conditions,

$$\overline{DP}_n = 1/(1 - p) ; \quad \overline{DP}_w = (1 + p)/(1 - p) \text{ and } \overline{DP}_w/\overline{DP}_n = (1 + p) .$$

For high-molecular weight polymers, i.e., for $1 - p \ll 1$, the ratio $\overline{DP}_w/\overline{DP}_n = 2$.

Other factors broadening molecular weight distribution deserve discussion. In some preparative methods involving living polymers, the monomer is slowly added to a well stirred reactor containing the required amount of an initiator. The polymerization continues as the monomer is supplied. Often, small traces of impurities "killing" the living polymers contaminate a monomer, and then the growth arising from the addition of that monomer takes place simultaneously with some irreversible termination caused by the impurity[229]. The resulting molecular weight distribution was discussed by Litt and Szwarc[223] and the results are shown graphically in Fig. 34.

The influence of a slow initiation on molecular weight distribution of a transferless and terminationless polymerization was treated by Nanda and Jains[224], who solved the equations

$$- d[M]/dt = k_i[M][I] + k_p[M]\sum[P_n^*]$$

$$d[P_n^*]/dt = k_p[M]([P_{n-1}^*] - [P_n^*]) ,$$

where M, I, and P_n^* denote a monomer, initiator, and a growing n-mer, respectively. The results were presented graphically in terms of the parameter $\zeta = (\overline{DP}_w/\overline{DP}_n) - 1$ as a function of $\ell n ([M]_0 - [M]_t)$ for various ratios k_p/k_i. The limiting values of $\overline{DP}_w$ and $\overline{DP}_n$ at infinite time were calculated by Litt[225], namely

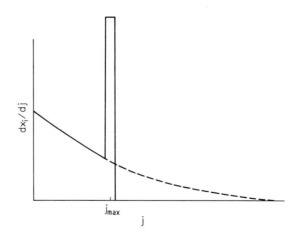

Fig. 34. Molecular weight distribution of polymers formed by continuous addition of monomer to living oligomers solution containing some terminating impurities

$$\overline{DP}_n = (k_p/k_i)\{1 + [\ell n(1 - f)]/f\}$$

and

$$\overline{DP}_w = (k_p/k_i)\{[\ell n(1 - f)^2]/[2 + f + \ell n(1 - f)]\} ,$$

where $1 - f$ denotes the ratio $[I]_\infty/[I]_0$.

Treatment of molecular weight distribution in polymerization initiated instantaneously but involving a spontaneous chain transfer was reported by Kyner, Radoe and Wales[226] and by Nanda[227]. Both groups simplified the calculations by considering the degree of polymerization as a continuous variable and not a discrete one.

Nanda extended his treatment to a system involving simultaneously a slow initiation and a transfer. The exact solution of that problem, treating the degree of polymerization as a discrete variable, was reported by Nanda and Jain[228]. The following kinetic scheme was assumed:

Initiation	$I + M \rightarrow P_1^*$,	k_i ,
propagation	$P_n^* + M \rightarrow P_{n+1}^*$,	k_p ,
spontaneous transfer	$P_n^* \rightarrow P_n + I$,	k_1 ,
transfer to monomer	$P_n^* + M \rightarrow P_n + P_1^*$,	k_2 .

The general solution is most complex; numerical calculations are not feasible even in the limits of a high-molecular weight product. For a high-molecular weight product, $\overline{DP}_n$ is given by:

$$\overline{DP}_n = (Y\beta + 1)/\{1 - \alpha\beta[1 - Y/(1 + \alpha)]\}$$

and a much more complex expression is derived for $\overline{DP}_w$. The above symbols are defined as $\alpha = k_1/k_p M_0'$, $\beta = M_0'/I_0$, $Y = (M_0' - M_t)/M_0'$, and $M_0' = M_0 - I_0$. The resulting expressions are sufficiently simple to permit numerical calculations if the stationary state approximation is adopted. For this approximation the plot of $n[P_n]$ vs. $n/\overline{DP}_n$ is shown in Fig. 35, and of $R = \overline{DP}_w/\overline{DP}_n$ vs. Y in Fig. 36. The last plot illustrates the variation of the inhomogeneity of molecular weight distribution with the degree of conversion for some chosen ratios $\alpha = k_1/k_p M_0'$ and k_2/k_p.

In a recent paper, Largo-Cabrerizo and Guzman[231] discussed molecular weight distribution achieved in a system involving a transfer agent T. The following kinetic scheme was assumed:

$I + M \rightarrow P_1^*$	instantaneous initiation	
$P_n^* + M \rightarrow P_{n+1}^*$,	k_p	
$P_n^* + T \rightarrow P_n + P_1^*$,	k_{t2} .	

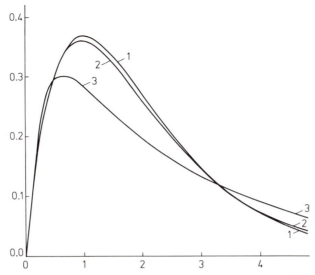

Fig. 35. Plot of weight fraction of n-mers, $n[P_n]$ vs. normalized degree of polymerization, $n/\overline{DP}_n$. *Curve (1) – spontaneous transfer, $\alpha = 6 \cdot 10^{-4}$; curves (2) and (3) – transfer to monomer*

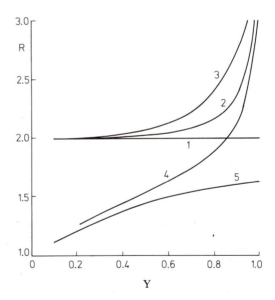

Fig. 36. Plot of the degree of inhomogeneity measured by $R = \overline{DP}_w/\overline{DP}_n$ vs. degree of conversion, Y, for various values of α and β

Three distinct cases were considered:
(a) The concentrations of a monomer and a transfer agent remain constant during the polymerization, this being achieved by a continuous supply of monomer and the transfer agent.
(b) Only the concentration of the transfer agent is kept constant while the concentration of the monomer decreases and tends to zero at $t = \infty$.

Table 3. The fraction of the residual transfer agent left after total depletion of the monomer for different values of $[M]_0/[T]_0$ and m

$[M]_0/[T]_0$	m = 1	5	20	50	200	1,000
1	0.500	0.699	0.859	0.924	0.974	0.993
10	0.091	0.456	0.767	0.883	0.963	0.991
50	0.020	0.331	0.708	0.855	0.955	0.989
100	0.010	0.288	0.684	0.843	0.952	0.089
500	0.002	0.209	0.631	0.817	0.941	0.987
1,000	0.001	0.182	0.610	0.805	0.941	0.986

(c)   Both concentrations, of the monomer and of the transfer agent, gradually decrease.

In the last case, provided that not all the transfer agent was consumed by the process, the ratio $[T]_\infty/[T]_0$ is given in Table 3 for various ratios $[M]_0/[T]_0$ and m $= (k_p - k_{t2})/k_{t2}$.

Another important factor affecting molecular weight distribution arises from reversibility of propagation. Its influence was discussed earlier in the section dealing with the thermodynamics of propagation, see p. 25 and Ref. 205.

Finally, some purely mechanical disturbances of polymerizing systems affect the molecular weight distribution of the product. Obviously, the inhomogeneity of temperature caused by inadequate stirring and slow heat transfer is a disturbing factor. The inhomogeneity of concentration is another. Figini[233] treated the effect of a dropwise addition of monomer to a well stirred reactor by assuming an exponential expansion of each droplet as it gets into the reactor. An alternative treatment was proposed by Litt[225], who visualized the mixing as a spreading of a droplet into a flat ribbon in a laminary rotating solution. On the whole, treatments of such problems are difficult and depend on the geometry of the system, the mode of stirring, and other conditions characterizing individual reactors.

Many studies of molecular weight distribution were carried out on samples prepared by anionic polymerization performed in the flow systems described by Schulz et al.[265, 366]. It is essential to ensure a turbulent character of flow when such equipment is used in the preparation of polymers of narrow molecular weight distribution; otherwise, some broadening of the distribution is observed[359, 367].

Finally, let it be stressed that for *irreversible* propagation the so-called "seeding" technique cannot improve molecular weight distribution of the product resulting from slower initiation than propagation. Since both processes are first order in monomer, their relative rates remain unaffected by variation of monomer concentration. Therefore, the same distribution is obtained on addition of small amounts of monomer to the initiator followed, after a waiting period, by the addition of the remainder, as by addition of all the monomer, provided that the mixing is adequate. On the other hand, reversibility of propagation allows a quantitative utilization of a sluggish initiator on addition of only a small fraction of a required monomer, provided that the waiting period is long compared with the rate of depropagation. Under such conditions the molecular weight distribution could be improved, i.e. it becomes sharper.

IV.3. Determination of Concentration of Living Polymers

Since the end-groups of ionically growing polymers may exist in a variety of states, the meaning of the term "concentration of living polymers" calls for clarification. What is meant is the *total* concentration of all the inter-convertible species which ultimately participate in the growth, even if momentarily only a small fraction of them is active. Polymers endowed with inert end-groups that neither propagate *nor can spontaneously be converted* into growth sustaining species are referred to as dead polymers.

Concentration of living polymers is determined in a relatively simple way when the equilibrium between the various inter-convertible species is rapidly established and their proportions in the investigated solution remain unaffected by polymerization and by their total concentration. In such a case, a variety of observable properties of living polymers, e.g. their electronic, IR, or magnetic spectra, could be used for the analysis. Whenever a concentration of colored, living polymers is determined from the intensity of an absorption spectrum, it is imperative to ascertain that neither dead polymers nor any other species present in the system under investigation absorbs in the chosen spectral region. For example, some irreversible reactions could convert the colored growing carbanionic end-groups into still colored but inert ones. The intensity of such an absorption spectrum, say at its λ_{max}, is then misleading. Fortunately, this kind of distortion of an observed spectrum is easily detected by comparing its shape with that of a genuine living polymer. Another phenomenon distorts the determination of concentration of living polymer measured by the height of a judiciously chosen peak in their magnetic spectrum. The pertinent signal could arise from averaging of signals of some interconvertible species. Its height could then be affected by the rate of interconversion which, in turn, might depend on the total concentration of living polymers.

In many systems most living polymers are in a dormant form and only a small but active fraction of them contributes to propagation. For example, more than 95% of living polymers could form the rather unreactive ion-pairs, while a low percentage is present as the highly reactive free ions that are responsible for the observed propagation. Similarly, aggregated lithium polystyrene salt is inert in hydrocarbon medium, and only a small fraction of the non-aggregated salt propagates the observed polymerization. The measured spectral properties are those of the dormant species, the active one contributing only a little. Nevertheless, the intensity of the spectra provide an acceptable measure of the concentration of all the living polymers, provided that the percentage of the active species is low.

Frequently the spectral features of the different species participating in propagation are closely similar. Whenever this is the case, variation of their relative concentrations leaves the spectrum virtually unaltered. This is often implicitly assumed. However, it is important to realize that examples to the contrary are known. In fact, the differences of spectral features of different species were effectively utilized in establishing their nature and proving their existence in the investigated systems (see, e.g.[280]).

In conclusion, the spectral methods are most useful and reliable if proper care is exerted in their applications. They are invaluable whenever it is desired to continuously monitor the concentration of living polymers. Partial destruction of living polymers, or any other changes in their nature taking place in the course of polymerization are then readily detected, especially when the absorbance is measured over a sufficiently wide spectral range.

Some irreversible reactions quantitatively convert living polymers, but not dead ones, into unique products, detection of which is easily and quantitatively achieved. It is immaterial whether such a reaction proceeds with only one of the inter-convertible species or with all of them, as long as the inter-conversion is fast. The concentration of the resulting product measures then the total concentration of living polymers, whatever their state. For example, addition of Michler ketone to solutions of living poly-styrene converts their carbanionic groups into an alcohol readily oxidized to an intensely colored dye. The conversion is quantitative and applicable for analytic usage[246].

Other examples of this "capping" technique were reported in the literature. For example, Saegusa[247] capped growing oxonium ions of polytetrahydrofuran and other polyethers with sodium phenolate, converting the end-group into phenolate. The concentration of the latter was determined by UV spectroscopy. Improvement of this method was reported by Barzikina et al.[257], who used picrates salts as capping agents.

Various titration techniques are based on the same principle. The agent used in titration has to react irreversibly and quantitatively with living polymers, but not with dead ones. The end-point is determined by a sharp change in some convenient property of the system, e.g. by the disappearance of the color of living polymers, provided that the dead polymers are colorless.

An unconventional technique was developed by Schulz and his co-workers[261]. It calls for determination of number average degree of polymerization, $\overline{DP}_n$, at various degrees

Table 4. Absorption maxima and extinction coefficients of living polymers, Na^+ counter-ion, THF solution

Living polymer derived from	λ_{max}/nm	$\varepsilon \times 10^{-4}$	Ref.
Styrene	342	1.2 (1.4)	a
α-Methylstyrene	352	1.2	b
Vinyl mesitylene	360	–	b
2-Vinyl pyridine	315	1.04	c
4-Vinyl pyridine	315	1.04	c
Vinyl biphenyl	405	–	i
1-Vinyl naphthalene	558	0.65	d
2-Vinyl naphthalene	410	0.91	d
9-Vinyl anthracene	392 and 372	–	j
1,1-Diphenyl ethylene	470	2.6	h
Butadiene (Li^+, cyclohexane)	275	0.83	e
Isoprene (Li^+, cyclohexane)	270	0.69	f
Methyl methacrylate (Li^+, THF)	335	0.24	g

a) D. N. Bhattacharyya, C. L. Lee, J. Smid, M. Szwarc: J. Phys. Chem., 69, 612 (1965)
b) D. N. Bhattacharyya, J. Smid, M. Szwarc: J. Polymer Sci., A, 3, 3099 (1965)
c) C. L. Lee, J. Smid, M. Szwarc: Trans. Faraday Soc., 59, 1192 (1963)
d) F. Bahsteter, J. Smid, M. Szwarc: J. Am. Chem. Soc., 85, 3909 (1963)
e) A. F. Johnson, D. J. Worsfold: J. Polymer Sci., A, 3, 449 (1965)
f) D. J. Worsfold, S. Bywater: Can. J. Chem., 42, 2884 (1964)
g) D. M. Wiles, S. Bywater: J. Polymer Sci., B, 2, 1175 (1964)
h) E. Ureta, J. Smid, M. Szwarc: J. Polymer Sci., A 1, 4, 2219 (1966)
i) A. Rembaum, J. Moacanin, E. Cuddihy, private communication.
j) A. Rembaum, A. Eisenberg, in: Macromolecular Reviews, Vol. 1, A. Peterlin, M. Goodman, S. Okamura, B. H. Zimm, H. F. Mark (Eds.), Intersc., New York, 1967

of conversion of the monomer into polymer. In the absence of termination and chain-transfer, the plot of the degree of conversion vs. $\overline{DP}_n$ is linear and its slope gives the concentration of living polymers. However, the authors reported[265,b)] that this method gives by 20–50% lower values for the concentrations of living polymers than titration with MeI or EtBr. On the other hand, a perfect correlation between the thus determined concentration of sodium polystyrene in tetrahydropyrane and the optical density of its solution was reported by Böhm and Schulz[300)], leading to molar absorbance of $\varepsilon = 1.5 \cdot 10^4 \pm 10\%$, in agreement with the findings of Bywater, $1.35 \cdot 10^4$ [279)] and Ivin, $1.3 \cdot 10^4$ [274)].

Finally, a quantitative conversion of initiator into living polymers is assumed in many studies. Provided that termination and other destructive reactions are absent, the initial concentration of the initiator is equal to the concentration of living polymers. Presence of a residual initiator in a non-quantitative initiation is readily detected.

Since optical spectroscopy is widely and successfully used in studies of many polymerizations involving carbanionic living polymers, the values of λ_{max} and of the respective molar absorbances of typical carbanionic living polymers are collected for the convenience of the reader in Tables 4 and 5.

The required molar absorbance, ε, of living polymers at a chosen wavelength, λ, is conveniently and reliably determined by adding to their solution a known amount of a "killing" reagent that quantitatively converts living polymers into dead ones. The concentration of the added reagent should be smaller than needed for their total destruction.

Table 5. Effects caused by the counter-ion and solvent on the absorption of living polystyrene

Solvent	Counter-ion	λ_{max}/nm	$\varepsilon \times 10^{-4}$	Ref.
THF	Li^+	337 (338)	1.00	a, b
Dioxane	Li^+	336	1.02	a
Benzene	Li^+	335	1.3	c
Cyclohexane	Li^+	328	1.35	d
THF	Na^+	342 (343)	1.20, 1.18, 1.4	a, b
Dioxane	Na^+	339	1.21	a
Benzene	Na^+	332	–	e
THF	K^+	343 (346)	1.20	a, b
Dioxane	K^+	340	1.21	a
Cyclohexane	K^+	330	1.32	–
THF	Rb^+	340	1.2	a
Dioxane	Rb^+	341	1.23	a
THF	Cs^+	345	1.3	a
Dioxane	Cs^+	342	1.24	a
Cyclohexane	Cs^+	333	1.25	–
THF	Ba^{2+}			f
THF	Sr^{2+}			g

a) D. N. Bhattacharyya, C. L. Lee, J. Smid, M. Szwarc: J. Phys. Chem., 69, 612 (1965)
b) S. Bywater, A. F. Johnson, D. J. Worsfold: Can. J. Chem., 42, 1255 (1964)
c) D. J. Worsfold, S. Bywater: Can. J. Chem., 38, 1891 (1960)
d) A. F. Johnson, D. J. Worsfold: J. Polymer Sci., A, 3, 449 (1965)
e) J. E. L. Roovers, S. Bywater: Trans. Faraday Soc., 62, 701 (1966)
f) B. De Groof, M. van Beylen, M. Szwarc: Macromolec. 8, 396 (1975)
g) C. De Smedt and M. van Beylen: ACS symposium Ser. 166, 127 (1981)

Then the decrease in the optical density of living polymers, in conjunction with the known initial concentration of the added reagent, allows one to determine the desired ε, provided that the dead polymers, the "killing" reagent, and the product of its reaction do not absorb at the wavelength chosen for the analysis. Of course, it is advisable to measure the optical density at λ_{max} of the absorbance in the visible, or near visible UV region. A comparison of the shape of the spectra recorded before and after the partial killing provides a valuable test of the reliability of that method. The most useful reagents of that kind are the high molecular weight alcohols, e.g. cetyl alcohol.

IV.4. Anionic Polymerization of Styrene and α-Methyl Styrene in Ethereal Solvents. Free Ions and Ion-Pairs

Kinetics of propagation of living polystyrene and poly-α-methyl styrene in various ethereal solvents were extensively investigated by several research groups. The results were revealing and of considerable importance since they led to understanding of the fundamental features of all ionic polymerizations and especially to a recognition of the nature and role of a variety of species participating in ionic propagation.

Worsfold and Bywater[270] were the first to apply the living polymer technique for determination of the absolute propagation constant of living poly-α-methyl styrene. Relative slowness of its polymerization in tetrahydrofuran allows the use of dilatometric technique and facilitates the investigation. The reaction was initiated by sodium naphthalenide and studied over temperatures ranging from $-70°$ to $-10\,°C$, the concentration of living polymers being varied from $2-30 \cdot 10^{-4}$ M. Plots of $\ell n(M–M_e)$ v. time were linear*, yielding the pseudo-first order rate constants. A plot of the latter constants vs. living polymers concentrations resulted in a straight line with a small positive intercept. The intercept is indicative of participation of free ions and ion-pairs in the propagation (see p. 92), although this was not realized at that time.

Polymerization of sodium polystyryl in dioxane was reported by Allen, Gee, and Stretch.[271]. The reaction, initiated by sodium naphthalenide, was first order in monomer and in living polymers, i.e. the calculated second order propagation constant was independent of the concentration of living polymers, being about 5 $M^{-1}s^{-1}$ at 25 °C. The activation energy, calculated from the data obtained in the range of 11° to 40 °C, was found to be 9 ± 3 kcal/mol, corresponding to frequency factor of $\sim 10^7$ $M^{-1}s^{-1}$. The authors concluded that the reaction is propagated by ion-pairs.

The work of Allen et al. was extended simultaneously by Bhattacharyya et al.[272] and by Dainton, Ivin and their students[274]. Both groups investigated the effect of counter-ions on the propagation constant. Bhattacharyya et al. monitored the progress of the reaction by a spectrophotometric technique and confirmed the independence of the bimolecular propagation constant of the concentration of sodium polystyryl varied by a factor of 24. The absence of free ions was demonstrated by the lack of conductance of the

---

* Equilibrium concentration of α-methyl styrene is rather high and cannot be neglected. Hence, the pseudo-first order constants have to be derived from plots of $\ell n(M-M_e)$, and not $\ell n(M)$, vs. time

investigated solutions – a not surprising result in view of the very low dielectric constant of dioxane, about 2.2. The propagation constants were claimed to be 0.9 $M^{-1}s^{-1}$ for the Li^+ salt, 3.4 $M^{-1}s^{-1}$ for the Na^+, and 20, 21, and 24.5 $M^{-1}s^{-1}$ for the K^+, Rb^+ and Cs^+ salts, respectively. The monotonic increase of reactivity with the radius of cation was interpreted as evidence for the tight, contact nature of ion-pairs of those salts in dioxane. Apparently, dioxane molecules (DOX) are incapable of separating these pairs, although crystallinic complexes such as $LiBr, 2 DOX^{276)}$ or $NaBF_4, 2 DOX^{277)}$ are known. Presumably in these complexes the cations are externally solvated by dioxane.

Dainton, Ivin et al.[274] used a dilatometric technique in their study of polymerization of alkali salts of living polystyrene in dioxane. They reported bimolecular constants for the Na^+, K^+, and Rb^+ salts which were somewhat higher than those claimed by Bhattacharyya et al., but a slightly lower constant was found for the Cs^+ salt. The respective activation energies and the corresponding A factors were determined, although in a rather narrow temperature range of 11–40 °C. For sodium polystyryl a conventional A factor of $6 \cdot 10^7$ $M^{-1}s^{-1}$ was calculated, but those pertaining to the K^+, Rb^+ and Cs^+ salts were low, namely $\sim 10^6$, $5 \cdot 10^4$, and $4 \cdot 10^5$ $M^{-1}s^{-1}$, respectively, implying a more complex mechanism of propagation.

A most thorough study of polymerization of sodium polystyryl in dioxane was reported by Komiyama, Böhm and Schulz[275]. Their results, shown graphically in Fig. 37, lead to E = 10.5 kcal/mol and A = $2.5 \cdot 10^8$ $M^{-1}s^{-1}$.

All the available results are collected in Table 6. The agreement between the findings of the various groups is fair. In view of the meticulous work of Schulz' group, their data seem to be the most reliable.

The very low dielectric constant of dioxane of about 2.2, comparable to that of hydrocarbons, may induce some aggregation of ion-pairs in this solvent. Such an aggrega-

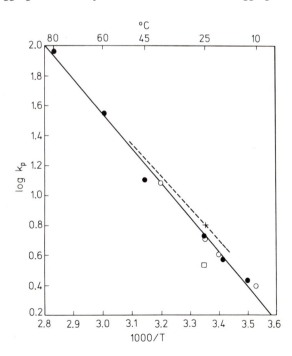

Fig. 37. Arrhenius plot of the propagation constant, k_p, for sodium polystyrene in dioxane. ● – the data of Schulz[275]; ○ – the data of Allen[271]; □ – the value from[272]; + – the data of Dainton[273]

Table 6. Bimolecular propagation constant, $k_p = k_\pm$, of polystyryl ion-pairs in dioxane at 25 °C

Cation	$k_p = k_\pm$ in $M^{-1}s^{-1}$;	E in kcal/mol;	A in s^{-1}	Ref.
Li^+	.9	–	–	272
Na^+	4.0–5.4	9 ± 3	$\sim 10^7$	271
Na^+	3.4	–	–	272
Na^+	6.5	9.5 ± 1	$6 \cdot 10^7$	274
Na^+	5.3–5.9	10.5	$2.5 \cdot 10^8$	275
K^+	20.	–	–	272
K^+	28 ± 1	6.1	$\sim 10^6$	274
Rb^+	21.5	–	–	272
Rb^+	34	4.4	$5 \cdot 10^4$	274
Cs^+	24.5	–	–	272
Cs^+	15 ± 5	6 ± 4	$4 \cdot 10^5$	274

tion is observed in hydrocarbon solutions of alkali salts of living polymers and leads to a more complex kinetics of propagation. However, since all the investigators mentioned above worked with polymers endowed with two active end-groups, the association, if any, would be intra-molecular. Such an association does not affect the kinetic character of the polymerization, although it should affect its rate. To clarify this issue, Shinohara[278] determined the propagation constant in dioxane of two samples of sodium polystyrene, one composed of polymers endowed with 2 active end-groups, while the polymers of the other had 1 end-group only. Both samples showed identical reactivities, implying lack of association in dioxane since an extremely high degree of inter-molecular association does not seem probable.

Polymerization of styrene initiated in cumyl-methyl ether by cumyl sodium is hardly affected by variation of concentration of living polymers or by the addition of sodium tetraphenyl boride[269]. This is not surprising since the dielectric constant of that solvent is also very low, 3.7 at 20 °C, although higher than of dioxane (2.2). The bimolecular propagation constant at $- 20$ °C is ~ 1 $M^{-1}s^{-1}$, a value expected for a propagation of contact ion-pairs.

The most far-reaching information about the nature of the species participating in anionic propagation was derived from studies of living polystyrene in tetrahydrofuran, THF. Kinetics of anionic propagation of sodium polystyryl in that solvent were first investigated by Geacentov et al.[263], who adopted a capillary flow technique for determining its rate. The expected behavior of living polymers was confirmed – in each run styrene consumption obeyed the first order kinetics. However, they noted a small but undeniable increase of the bimolecular rate constant, k_p,

$$k_p = (d\ell n[\text{styrene}]/dt)/[\text{living polymers}],$$

on decreasing the living polymer concentration, c_L. This phenomenon was thoroughly investigated by Szwarc and his co-workers[264] and simultaneously by Schulz and his students[265]. Both reported a linear dependence of k_p on reciprocal of the square-root concentration of living polymers, $1/c_L^{1/2}$, and rationalized it by invoking free polystyryl anions, $\sim\!\!\sim S^-$, and their ion-pairs, $\sim\!\!\sim S^-, Cat^+$, as the species responsible for propagation.

The free $\wedge\wedge\wedge$ S$^-$ ions seem to be much more reactive than their ion-pairs. Denoting by k_- and $k_\pm$ their respective propagation constants and by α the mole fraction of the free ions, or the free $\wedge\wedge\wedge$ S$^-$ end-groups*, we find the observed propagation constant, k_p, to be

$$k_p = \alpha k_- + (1 - \alpha)k_\pm ,$$

α denoting the mole fraction of free ions. Assuming the validity of the mass-law for the dissociation,

$$\wedge\wedge\wedge S^-, Cat^+ \rightleftharpoons \wedge\wedge\wedge S^- + Cat^+ , \quad K_{diss} ,$$

One deduces

$$\alpha^2 \cdot c_L/(1 - \alpha) = K_{diss} ,$$

and for low degrees of dissociation, i.e. $\alpha \ll 1$, the approximation

$$\alpha^2 = K_{diss}/c_L .$$

Within this approximation one gets

$$k_p = k_\pm + (k_- - k_\pm)K_{diss}^{1/2}/c_L^{1/2} ,$$

i.e. the relation accounting for the observed linear plot of k_p vs. $1/c_L^{1/2}$. Further simplification, justified whenever $k_\pm \ll k_-$, leads to the approximate relation

$$k_p = k_\pm + k_- K_{diss}^{1/2}/c_L^{1/2} .$$

Accepting this interpretation and the above approximations, one finds the intercept of a plot k_p vs. $1/c_L^{1/2}$ to be equal to $k_\pm$, and its slope to $k_- K_{diss}^{1/2}$.

Plots of k_p vs. $1/c_L^{1/2}$ shown in Fig. 38 lead to several valuable conclusions[264]. The slopes of the lines shown in that figure decrease as the cation varies from Li$^+$ to Na$^+$, K$^+$, Rb$^+$ and Cs$^+$. Since k_- is independent of the nature of the cation, the observed decrease of slopes signifies a decrease of K_{diss} along that series. This might appear strange – the Coulombic work needed to separate the oppositely charged ions decreases as the radia of cations increase on going from Li$^+$ through Na$^+$, K$^+$, Rb$^+$ to Cs$^+$. However, the solvation of cations has to be considered and its degree increases as the radia of cations decrease. It seems that the solvation energy of alkali cations by THF molecules is greater than the Coulombic work needed for ion-pair separation. Indeed, the conductance of THF solutions of alkali salts of living polystyrene increases on lowering their temperature[123, 303] in spite of their increasing viscosity. This manifests the exothermic character of the dissociation. A similar behavior was noted for the conductance of other alkali salts

---

* The polymers investigated by Szwarc and by Schulz resulted from electron-transfer initiation. Therefore, as pointed out in the early publication of Szwarc[16], each polymer was endowed with two active end-groups. This was experimentally confirmed by Barnikol and Schulz[261]

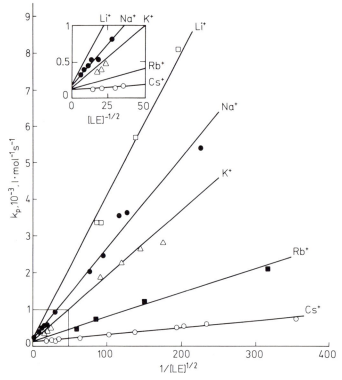

Fig. 38. Plots of the propagation constant k_p for salts of living polystyrene in THF vs. $1/C_L^{1/2}$. $\square$ – Li$^+$; $\bullet$ – Na$^+$, $\triangle$ – K$^+$; $\blacksquare$ – Rb$^+$; $\bigcirc$ – Cs$^+$. T = 25 °C

in THF[283]. Conductometric determination of the dissociation constants of alkali salts of living polystyrene, carried out over a wide temperature range[123], showed the exothermicity of the dissociation of sodium polystyrene to be about 8 kcal/mol*, whereas that of its cesium salt was only about 2 kcal/mol. These findings imply a stronger solvation by THF of Na$^+$ than Cs$^+$ – a conclusion corroborated by the determination of the respective Stokes radia[283, a]. The Stokes radius of Na$^+$ was found to be about 4 A, whereas that of Cs$^+$ was only ~ 2 A, i.e., in THF solution the smaller Na$^+$ is more bulky than the larger Cs$^+$.

Some words of caution are in place. Determination of the dissociation constants of ion-pairs is achieved by measuring the equivalent conductance of their solution, Λ, at decreasing concentrations of the salt, c. Following Fuoss[284], one plots F/Λ vs. $f^2 c \Lambda / F$, where F and f are functions of c, Λ and of the dielectric constant, viscosity, and temperature of the solvent given in Ref. 284. The intercept and the initial slope of the resulting plot extrapolated to 0 salt concentration give $1/\Lambda_0$ and $1/K_{diss} \Lambda_0^2$, respectively, where Λ_0 denotes the limiting conductance.

---

* A somewhat smaller value of ~ 6.5 kcal/mol is deduced from Ref. 303

To ensure a reliable extrapolation to c = 0 for scarcely dissociated pairs, i.e. for K_{diss} $\leqslant 10^{-7}$, it is imperative to extend the measurements to extremely low concentrations of the salt, at least down to 10^{-6} M. The K_{diss} values of the salts of living polystyrene in THF are of this order of magnitude. Since traces of moisture, alcohol, O_2, etc., destroy them, a special dilution procedure is needed to avoid introduction of any impurities that could vitiate the results. Such a procedure is described in Ref. 264.

Further difficulties are encountered in the determination of Λ_0's. The $1/\Lambda$'s are large for scarcely dissociated pairs; hence, the Fuoss lines are steep. Consequently, the intercepts of such lines, plotted in the appropriate scale, are too small to be measured. Therefore, some alternative methods are needed for their evaluation (see, e.g.[123, 285−287, 292]).

Propagation constant of the free polystyryl anion in THF, i.e. k_-, could be calculated from the kinetic data provided that the pertinent dissociation constant is known. In view of the previously discussed difficulties associated with the conductance determination of K_{diss}, it is desired to develop some alternative method that could provide the values of k_- and K_{diss}. Such a method was outlined by Bhattacharyya et al.[264] and simultaneously by Schulz et al.[265, c)]. Both teams determined the observed propagation constant, k_p, in the presence of a readily dissociated salt sharing a common cation with ion-pairs of the living polymer. The system: Sodium salt of living polystyrene – Na^+, BPh_4^- – THF, provides an example. The fraction α of the dissociated living polymer at its total concentration c_L is given by

$$\alpha = K_{diss}/(K_{diss} + [Cat^+])$$

where K_{diss} is the dissociation constant of ion-pairs of living polymers and $[Cat^+] = x$ denotes the concentration of the common cations formed through the dissociation of ion-pairs of the living polymers and of the added salt. The value of x is calculated by solving three simultaneous equations,

$$x = \alpha c_L + \beta c_A ,$$

$$\alpha = K_{diss}/(K_{diss} + x) ,$$

and

$$\beta = K_{diss, A}/(K_{diss, A} + x) , \quad .$$

where c_A, β, and $K_{diss, A}$ denote the concentration, degree of dissociation, and the dissociation constant, respectively, of the added salt. By a judicious choice of the salt and its concentration, x could be made much larger than K_{diss} and then the following approximate relation applies:

$$x^3 + K_{diss}x^2 - (c_L K_{diss} + c_A K_{diss, A})x - c_L K_{diss} \cdot K_{diss, A} = 0$$

with $\alpha \approx K_{diss}/x$. Under those conditions, and for $k_\pm \ll k_-$, the observed propagation constant, k_p, is given by

$$k_p = k_\pm + k_- \cdot K_{diss}/x ,$$

i.e. a plot of k_p vs. $1/x$ is linear with intercept of $k_\pm$ and slope of k_-K_{diss}. Such a plot is shown in Fig. 39 for polymerization of the sodium salt of living polystyrene in the presence of a variable concentration of NaBPh$_4$. Let its slope be denoted by s', while the slope of the line resulting from plotting k_p determined in the absence of the added salt vs. $1/c_L^{1/2}$ is denoted by S. Since s' $= k_-K_{diss}$ and S $= k_- \cdot K_{diss}^{1/2}$, it follows that

$$K_{diss} = (S/s')^2 \quad \text{and} \quad k_- = S^2/s' \ .$$

By this approach Bhattacharyya et al.[264] and Schmitt and Schulz[289] determined the relevant K_{diss} of sodium polystyryl in THF and the absolute propagation constant of the free polystyryl anions.

Experiments performed with other alkali salts of living polystyrene and the corresponding tetraphenyl borides led to the results collected in the second column of Table 7. The last column of that table gives the K_{diss} calculated from the conductance data (slopes of Fuoss' lines) in conjunction with the respective Λ_0. The dissociation constant of sodium polystyryl first reported by Worsfold and Bywater[285], served as a reference point. The agreement between both sets of data is satisfactory.

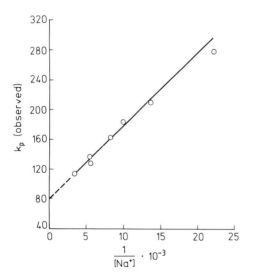

Fig. 39. Plot of the propagation constant k_p of sodium polystyrene in THF as function of $1/[Na^+]$. The concentration of Na$^+$ cations varied by addition of NaBPh$_4$. T = 25 °C

Table 7. The dissociation constants of alkali salts of living polystyryl in THF at 25 °C[264]

Cation	$10^7 K_{diss, Cat+}$ /M, kinetic	$10^7 K_{diss, Cat+}$ /M, conductance
Li$^+$	2.2	1.9
Na$^+$	1.5	1.5 (from[285])[a]
K$^+$	0.77	0.7
Rb$^+$	0.10	–
Cs$^+$	0.02	0.03

a A lower value of $0.8 \cdot 10^{-7}$ M was reported by Böhm and Schulz[304]

Similar investigations were carried out using other ethereal solvents. Propagation of living polystyrene in tetrahydropyrane, THP, is substantially slower than in THF – a not surprising observation since the dielectric constant of the former ether (5.5) is lower than that of the latter (7.8). Dilatometric[273] and batch[267, a] techniques were used in following its course because this polymerization is relatively slow. The observed bimolecular constants increase on dilution, implying again that free ions and ion-pairs partipate in the growth process. The early work of Barnikol and Schulz[267], extended by a thorough study of Böhm and Schulz[300], was limited to the sodium salts. The constants $k_-K_{diss}^{1/2}$ were determined for temperatures ranging from $-40°$ to $50\,°C$, the results being remarkably similar to those reported by Van Beylen et al.[299].

T°C	-40,	-20,	0,	25	Ref.
$k_-K_{diss}^{1/2}M^{1/2}s^1$	0.37,	0.47,	0.63,	1.02	[300] Table 1
$k_-K_{diss}^{1/2}M^{1/2}s^1$	0.27,	0.47,	0.63,	1.01	[299] Table 4.

By using the previously discussed kinetic approach, that yields k_-K_{diss} from the dependence of k_p on $1/[Na^+]$, Böhm and Schulz[300] determined K_{diss} and k_- for sodium polystyrene in THP. As expected, the K_{diss} is much lower in THP, $\sim 10^{-10}$ M, than in THF, but the values of k_- are similar in both systems. The value of the dissociation constant of $NaBPh_4$ in THP, needed for the calculation, was found to be $2.2 \cdot 10^{-6}$ M^{279}.

Propagation of other alkali salts of polystyrene in THP were studied by Van Beylen et al.[299] for the Li^+ and Na^+ salts, and by Bywater's group[279] for those of Li^+, Na^+, K^+, and Rb^+. Both groups directly determined the respective K_{diss} by conductance studies. The exothermicity of sodium polystyryl dissociation in THP is substantially lower than in THF, ~ 3 kcal/mol[299] or 2.5 kcal/mol[300], compared with ≈ 8 kcal/mol found for the dissociation in THF[264]. Still lower $-\Delta H_{diss}$ is found for the dissociation of the Li^+ salt[299]. It should be stressed, however, that all the pertinent van't Hoff plots are curved, namely $-\Delta H_{diss}$ decreases at lower temperatures. The significance of this observation is discussed later.

Propagation constants, $k_\pm$, of the polystyryl ion-pairs in THP were derived, like in the other studies, by extrapolating the plots of k_p vs. $1/c_L^{1/2}$ to infinite concentration of living polymers, or by suppression of the dissociation of ion-pairs of living polymers caused by the addition of appropriate tetraphenyl boride salts. The kinetic and conductometric results of all of these studies, summarized in Table 8, show a fair agreement between the data reported by all the investigators.

Propagation of alkali salts of living polystyrene in 2-Me-THF was investigated by Van Beylen et al.[299]. Participation of free ions and their pairs in the growth process taking place in that solvent is undeniable; however, the reported data should be treated as tentative only. Some spontaneous destruction of living polymers occurs in that solvent, and this complicates the kinetic studies. Since various purification methods did not eliminate these perturbing effects, it was suggested that living polymers are terminated by proton abstraction from the methyl group of the solvent.

Sodium polystyryl is also unstable in 3-Me-THF[302]. Kinetics of propagation was studied in this solvent over the most extensive temperature range, down to $-110\,°C$. The values of $k_-K_{diss}^{1/2} = 8.8$ $M^{-1/2}s^{-1}$ and of $k_-K_{diss} = 5.2 \cdot 10^{-4}$ s^{-1} were derived from the slopes of k_p vs. $1/c_L^{1/2}$ and vs. $1/[Na^+]$, respectively, both slopes determined for $20\,°C$.

Table 8. Propagation of alkali salts of living polystyrene in tetrahydropyrane (THP) at $\sim 25\,°C$

Cation	$k_- K_{diss}^{1/2}$ $M^{1/2}s$	$10^{10}K_{diss}$ M^{-1};	$k_\pm$ $M s$	Ref.
Li$^+$	2.2	1.9	~ 19	279
Li$^+$	1.1	2.5	< 10	299
Na$^+$	1.1	–	~ 7	267
Na$^+$	1.0	2.1	$< 5^a$	299
Na$^+$	–	–	14	268
Na$^+$	1.7	1.7	17	279
Na$^+$	1.0	0.7	14	300
K$^+$	–	–	73	268
K$^+$	2.8	4.0	30 ± 5	279
Rb$^+$	–	–	83	268
Rb$^+$	1.9	3.7	40 ± 5	279
Cs$^+$	–	–	53	268

^{a} Significantly, $k_\pm$ was found to be ≈ 12 $M^{-1}s^{-1}$ when the experiments were performed in the presence of NaBPh$_4$ (see Fig. 3 of Ref. 326)

Hence, $K_{diss} = 4.5 \cdot 10^{-9}$ M and $k_- = 1.5 \cdot 10^5$ $M^{-1}s^{-1}$. The required dissociation constant of NaBPh$_4$ in 3-Me-THF was determined from the conductance of its solution as $1.1 \cdot 10^{-5}$ M. Direct conductance studies of sodium polystyryl in 3-Me-THF led to $K_{diss} = 7.8 \cdot 10^{-9}$ M. A possible reason for the discrepancy between kinetic and conductance findings was discussed in that paper and in Ref.[289 b]. The interesting temperature dependence of $k_\pm$ will be discussed later.

Participation of free anions and their ion-pairs in propagation of sodium polystyryl in oxepane was demonstrated by Löhr and Bywater[301]. The dielectric constant of this ether is only by about 10% lower that of THP; nevertheless, the dissociation constant of sodium polystyryl in oxepane, viz. $7.3 \cdot 10^{-12}$ M at 30 °C, is substantially lower than that found in THP. The value of $k_- K_{diss}^{1/2}$ was reported as 0.27 $M^{-1/2}s^{-1}$ at 30 °C, yielding $k_- = 1 \cdot 10^5$ $M^{-1}s^{-1}$. The significance of the value of the respective $k_\pm = 14$ $M^{-1}s^{-1}$ and its temperature dependence will be discussed later.

Kinetic and conductance investigations of the sodium and cesium salts of living polystyrene in dimethoxyethane, DME, were first reported by Shimomura, Smid, and Szwarc[282]. The kinetic data had to be extrapolated to 0 time due to the instability of the sodium salt in that solvent. For the same reason a direct conductance study was prevented. The cesium salt, prepared by initiating the polymerization with cumyl cesium, is more stable in DME than the sodium salt; therefore, the pertinent kinetic observations and conductance studies were readily accomplished.

In spite of these difficulties, two important conclusions were derived from those studies. The polymerization of sodium polystyryl is by an order of magnitude faster in DME than in THF, implying a substantially greater degree of dissociation of the respective ion-pairs into free ions. The propagation constant, $k_\pm$, is about 50 times larger than that determined in THF and $\sim 1,000$ times larger than the $k_\pm$ in dioxane. The temperature dependence of $k_\pm$ of the sodium salt in DME is given by a curve with a maximum at about 20 °C. The behavior of cesium salt is more conventional.

The findings of Shimomura et al.[282] were soon confirmed by Schulz' group. Their more precise work[303, 306] allowed a direct determination of the dissociation constant of sodium polystyryl in DME over a wide temperature range. They reported the $k_\pm$ values only by $\sim 20\%$ higher than those found by Shimomura and fully confirmed the unconventional temperature dependence of $k_\pm$. The high degree of dissociation of sodium polystyryl ion-pairs in DME, comparable to that of $NaBPh_4^{[283\,a]}$, demanded the application of the non-approximated equation,

$$k_p = k_\pm + (k_- - k_\pm)K_{diss}/([Na^+] + K_{diss}) ,$$

in evaluation of the kinetic data derived from the dependence of k_p on the concentration of sodium ions. The complex equations leading to the desired constants were solved with the help of a computer minimizing program[311].

The high values of $k_\pm$ of sodium polystyryl in DME and its unconventional temperature dependence, as well as the clearly demonstrated negative temperature dependence of $k_\pm$ of sodium polystyryl in THF[281], led to significant conclusions discussed in the next section.

Propagation of alkali salts of living poly-α-methyl styrene has been less extensively investigated than of living polystyrene. Following the pioneering work of Bywater and Worsfold[270], Dainton and Ivin studied the propagation of various alkali salts of living poly-α-methyl styrene carried out in THF, THP, and dioxane[273, 274, 312]. Since the ceiling temperature for these systems is low, propagation, as well as depropagation, were investigated. Notably, the rate constants derived from the depropagation experiments were consistently by about 15% higher than those deduced from the propagation runs. This divergence was attributed to the different initial conditions prevailing in the respective experiments[274].

The behavior of sodium and potassium salts of poly-α-methyl styryl in THF resembles that of the corresponding polystyryl salts. The plots of the apparent bimolecular constants k_p vs. $1/C_L^{1/2}$ were linear. From their slopes and intercepts the $k_- K_{diss}^{1/2}$ and $k_\pm$ values were deduced, namely 0.03 $M^{-1/2}s^{-1}$ and 0.014 $M^{-1/2}s^{-1}$ for the $k_- K_{diss}^{1/2}$ of the sodium salt at $-31\,°C$ and $-59\,°C$, respectively, and 0.6 $M^{-1}s^{-1}$ and 0.42 $M^{-1}s^{-1}$ for the $k_\pm$ determined at the same temperatures. The corresponding data obtained for the potassium salt led to $k_- K_{diss}^{1/2} = 0.0022\ M^{-1/2}s^{-1}$ and $k_\pm = 0.012\ M^{-1}s^{-1}$ at $-59\,°C$. Like in the case of living polystyryl, the potassium salt is less dissociated and its pairs are less reactive than of the sodium salt.

The k_- values were found to be 49 $M^{-1}s^{-1}$ at $-31\,°C$ and 9 $M^{-1}s^{-1}$ at $-59\,°C$, by combining the kinetic data with the values of K_{diss} directly determined through conductance measurements[313]. The $k_- = 7\ M^{-1}s^{-1}$ at $-59\,°C$, derived from the data obtained for the potassium salt, is consistent with the previous finding. The respective activation energy and the frequency factor were reported as 7 kcal/mol and $10^8\ M^{-1}s^{-1}$.

The slowness of α-methyl-styrene homo-polymerization is attributed to the steric strain caused by the presence of bulky methyl groups. Such a strain is evident in equilibrium between the living poly-α-methyl styrene and its monomer. The inductive effect of the methyl group makes the monomer less reactive than styrene[314] but it enhances the reactivity of the poly-α-methyl-styrene carbanions. Indeed, living poly-α-methyl-styrene oligomers are efficient initiators of anionic polymerization of styrene.

The effect of cations on the reactivity of poly-α-methyl-styrene ion-pairs was the main subject of studies by Dainton and Ivin[312, 273]. The results reported for the propagation of Na$^+$, K$^+$, Rb$^+$, and Cs$^+$ salts in tetrahydropyrane are presented graphically in Fig. 40, and those in dioxane in Fig. 41. In those solvents the sodium ion-pairs are substantially less reactive than the K$^+$, Rb$^+$ and Cs$^+$ pairs, an order of reactivity resembling that observed for the alkali salts of living polystyryl (see Tables 6 and 8, p. 92 and 98). However, the reactivity of lithium ion-pairs in THP is abnormally high, $k_\pm = 2.6 \, M^{-1}s^{-1}$ at 25 °C. Spectroscopic evidence[307] suggests a change in the nature of the ion-pair, but some rapidly occurring reactions converting the living into dead polymers make such conclusions questionable.

The exceptional behavior of living α-methyl-styrene dimer was reported by Lee et al.[192], who utilized a stirred-flow technique in studying its reversible conversion into trimer. The conductance characteristic of these oligomers was investigated by Comyn et al.[307]. Some results pertaining to kinetics of addition of α-methyl styrene to its living tetramer were reported in[84]. Unfortunately, the difference between free ions and ion-pairs was not yet appreciated at the time of these investigations.

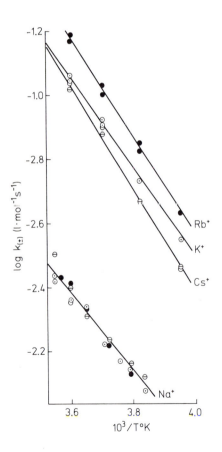

Fig. 40. Arrhenius plots of the propagation constant of ion-pairs, $k_\pm$, of poly-α-methylstyrene salts in tetrahydrofurane for Rb$^+$, K$^+$, Cs$^+$, Na$^+$

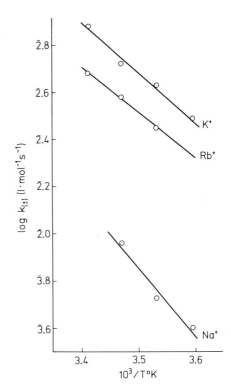

Fig. 41. Arrhenius plots of the propagation constant of ion-pairs, $k_{\pm}$, of poly-α-methylstyrene salts in dioxane for K^+, Rb^+, Na^+

IV.5. Temperature Dependence. Various Kinds of Ion-Pairs

The results discussed so far reveal the participation of free anions and their cation-anion pairs in propagation of anionic polymerization. In the case of styrene, the reactivity of free carbanions is only slightly affected by the nature of ethereal solvents in which the reaction proceeds. In fact, the results of Schulz' group demonstrate a nearly identical reactivity of the respective carbanions in THF, THP, DME, 3-Me-THF, oxepane, and HMPA (hexamethyl-phosphoric-triamide). As shown in Fig. 42, all the results fit a single, straight Arrhenius plot over a wide temperature range of $-50\,°C$ to $50\,°C$, with maximum deviation amounting only to $\sim 25\%$. Even if the divergence in the free anion's reactivity had been larger than claimed by Schulz, there is little doubt that the variation in polystyryl anions' reactivity is not larger than a factor of 2.

In contrast, the reactivities of ion-pairs vary enormously, especially on changing the solvent. For example, the propagation constant of sodium polystyryl ion-pairs at ambient temperature increases from about 5 $M^{-1}s^{-1}$ in dioxane to about 5,000 $M^{-1}s^{-1}$ in dimethoxyethane, raising the question of whether the label "ion-pair" is sufficient to describe these entities.

The problem of different kinds of ion-pairs was clearly recognized in 1954 in the publications by Fuoss[317] and by Winstein[318]. The approaches of these two independent-

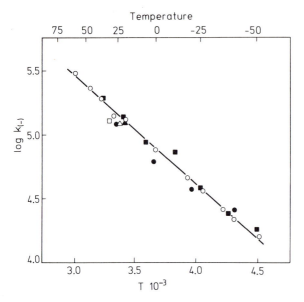

Fig. 42. Arrhenius plot of the propagation constant of free polystyryl ions in various solvents. △ – HMPA; ■ – DME, ○ – THF, ▲ – 3-MeTHF, ● – THP, ☐ – oxepane. Solid line: $E_- = 4$ kcal/mol; $A = 10^8 \, M^{-1}s^{-1}$

groups were vastly different; for our purpose Fuoss' approach is preferred. In solvents ions are surrounded by a "solvation shell" – a partially organized shell of solvent molecules influenced by the electric field generated by the ion. Small ions, i.e. Li^+, Na^+, etc., may generate very tight shells in appropriate solvents that retain their integrity during their Brownian walk. On colliding with counter-ions, they may form relatively long-lived pairs, their solvation shells being retained, although distorted by the interaction with the gegen-ions. In such a process, loose, solvent separated pairs are formed. However, at some instant the solvent molecules separating the ions of a loose pair may be squeezed out in a process yielding tight, "contact" pairs. The poorly solvated ions could directly form tight pairs on their collisons with counter-ions without previous formation of loose pairs.

In many systems the alternative forms discussed here correspond to two relatively deep minima in the appropriate potential energy hyper-surface. Thence, it is justified to treat these two kinds of pairs as two thermodynamically distinct species that are in equilibrium with each other. Each of these species possesses specific properties, like reactivities, which could be vastly different[288, 323].

Alternatively, in other systems the minima in the potential energy hyper-surface may be shallow and temperature dependent. In such a case, the concept of two distinct kinds of ion-pairs loses its usefulness[319, 324].

More complex systems are also known and then it is necessary to differentiate between more than two kinds of pairs. For example, in a solvent containing a small amount of a powerful solvating agent, X, three kinds of ion-pairs were observed[319, 320]: the ordinary tight pairs surrounded by solvent molecules, pairs externally associated with X, and pairs separated by X. Such systems are governed by the equilibria:

$$A^-, Cat^+ + X \rightleftharpoons A^-, Cat^+, X \rightleftharpoons A^-, X, Cat^+ .$$

Are the labels "tight pairs" and "loose pairs" sufficient to describe the species under consideration? For example, is a loose pair separated by DME molecules identical with a pair separated by THF molecules? Surely, on a precise examination, they should reveal some differences. However, provided that the effect of changeable solvent on properties of each kind of pairs is small, while the difference in their properties, e.g. reactivities, is large, then further distinctions may be superfluous.

In the early papers describing anionic polymerization of sodium polystyryl it was suggested that two kinds of pairs contribute to their propagation[264(b), 265, 267, 272]. The first compelling evidence revealing the substantially higher reactivity of loose sodium polystyryl pairs than of tight ones was furnished by the negative temperature dependence of $k_\pm$ observed in the polymerization of sodium polystyryl in THF[123], and by the maximum in the Arrhenius plot of $k_\pm$ determined for the polymerization of that salt in DME[282]. These results, depicted in Fig. 43, imply a high reactivity of the scarce, loose pairs and low reactivity of the abundant tight pairs*.

Conversion of tight into loose pairs is usually exothermic, e.g. for

$$(\text{Polystyryl}^-, \text{Na}^+)_{tight} \rightleftharpoons (\text{Polystyryl}^-, //, \text{Na}^+)_{loose} \quad \Delta H_{t,\ell} < 0 \, .$$

The observed $k_\pm$ is given by

$$k_\pm = ak_\ell + (1 - a)k_t = (K_{t,\ell} \cdot k_\ell + k_t)/(1 + K_{t,\ell})$$

where k_t and k_ℓ denote the propagation constants of tight and loose pairs, respectively, $K_{t,\ell}$ is the equilibrium constant of their inter-conversion, and a is the mole fraction of

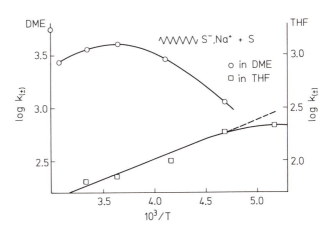

Fig. 43. Arrhenius plots of the propagation constant, $k_\pm$, of sodium polystyrene ion-pairs in DME (*upper curve*) and THF (*lower curve*). Note the "negative activation energy" for the propagation in THF and the maximum for the propagation in DME

---

* This is not a general statement. It applies to anionic polymerization of styrene and similar monomers, but examples to the contrary are known and will be discussed later

loose ion-pairs. Temperature dependence of $k_\pm$ is determined by the genuine activation energies E_t and E_ℓ of propagation by the tight and loose pairs, and by the $\Delta H_{t,\ell}$. For $k_\ell K_{t,\ell} \gg k_t$ but $K_{t,\ell} \ll 1$ the approximation

$$k_\pm = k_\ell K_{t,\ell}$$

is valid, and then a negative "activation energy" is observed provided that $-\Delta H_{t,\ell} > E_\ell$. These conditions are fulfilled in the polymerization of sodium polystyryl proceeding in THF at temperatures ranging from 25 °C to ~ -50 °C. However, at some still lower temperatures the Arrhenius plot should go through maximum, since α approaches unity as temperatures decreases and then the temperature dependence of $k_\pm$ is governed only by E_ℓ. Indeed, this behavior was observed for the sodium polystyryl polymerization taking place in DME, as shown by Fig. 43. The maximum appears at a not too low temperature because $K_{t,\ell}$, the equilibrium constant of inter-conversion of tight into loose pairs in DME, is high even at ambient temperature.

Schulz stressed[289] that at some sufficiently high temperatures the value of $k_\ell K_{t,\ell}$ becomes smaller than k_t and then the observed propagation constant, $k_\pm$ is mainly affected by k_t. Hence, the Arrhenius plot has to go through a minimum in this temperature range and eventually $k_\pm$ would rise with rising temperature due to activation energy associated with the propagation of tight ion-pairs. Indeed, S-shaped Arrhenius plots, shown in Fig. 44, were reported for sodium polystyryl polymerization proceeding in THF and 3-Me-THF when the reaction was investigated over a very wide range of temperatures.

The thorough and most extensive studies of sodium polystyryl polymerization reported by Schulz' group and summarized in their two reviews[322, 305] led to the determination of six parameters: the activation energies, E_t and E_ℓ and the corresponding frequency factors, A_t and A_ℓ of propagation by tight and loose pairs, respectively, and the heat, $\Delta H_{t,\ell}$, and entropies, $\Delta S_{t,\ell}$, of their inter-conversion. These parameters were obtained by computer fitting of the curves given by

$$\ell nk_\pm = \{A_\ell exp(-E_\ell - \Delta H_{t,\ell} + T\Delta S_{t,\ell})/RT + A_t exp(-E_t/RT)\}/$$
$$\{1 + exp(-\Delta H_{t,\ell} + T\Delta S_{t,\ell})/RT\}$$

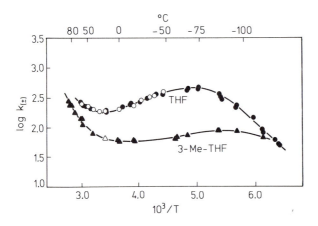

Fig. 44. Arrhenius plots of the propagation constant, $k_\pm$, of sodium polystyrene ion-pairs in THF and 3-Me-THF. Note the sigmoidal shape of the curves

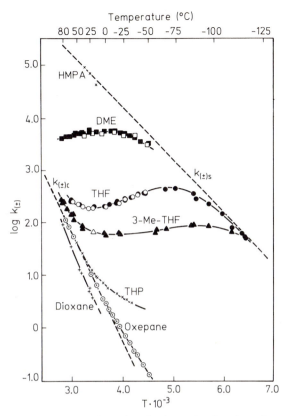

Fig. 45. Arrhenius plots of the propagation constants of ion-pairs of polystyryl sodium in various solvents. The curves approach a common assymptote at low temperatures, interpreted as a linear Arrhenius plot referring to the propagation constant of loose ion-pairs, and the curves approach again a common assymptote at high temperatures, interpreted as a linear Arrhenius plot referring to the propagation constant of tight ion-pairs

to the $\ell nk_\pm - 1/T$ data-points obtained for each solvent system over a sufficiently wide range of temperatures. Furthermore, it was tacitly assumed that A_ℓ, E_ℓ, A_t, E_t, $\Delta H_{t,\ell}$, and $\Delta S_{t,\ell}$ are all *temperature independent*. The results obtained for four solvent systems, DME, THF, 3-Me-THF, and THP, are given in Tables 9 and 10, and are graphically presented in Fig. 45. Their self-consistency is impressive, and moreover, the derived A_t's and E_t's favorably compare with those obtained from studies of the oxepane and dioxane system. In the latter solvents, the contribution of loose pairs to the propagation is negligible, preventing therefore the determination of the respective A_ℓ, E_ℓ, $\Delta H_{t,\ell}$, and $\Delta S_{t,\ell}$ but facilitating the determination of E_t and A_t.

The kinetic results discussed in the preceding paragraphs were supplemented by the data derived from conductance studies, namely those yielding the pertinent dissociation constants of ion-pairs, K_{diss}. Plots of ℓnK_{diss} *vs.* $1/T$ are curved for many salt-ethereal solvent systems, as exemplified by Fig. 46. At very low temperatures they are flat but become steeper as the temperature rises[281, 282, 288, 303, 309, 324, etc.]. This behavior is accounted for by the following equilibria[283]:

Table 9. The absolute values of k_t and k_ℓ at 25 °C and their Arrhenius parameters for anionic polymerization of sodium polystyryl in various ethereal solvents

Solvent	k_t $M^{-1}s^{-1}$	$\log A_t$	E_t kcal/mol	$10^{-4}k_\ell$ $M^{-1}s^{-1}$	$\log A_\ell$	E_ℓ kcal/mol	Ref.
DME	12.5	7.8	9.2	5.5	7.8	4.2	303, 306
THF	34.	7.8	8.6	2.4	8.3	4.7	289 a
3-Me-THF	20	8.0	9.2	12.4	8.3	4.4	289 b
THP	10.7	8.1	9.7	5.3	8.0	4.5	300
Oxepane	18	8.2	9.8	–	–	–	301
Dioxane	5.5	8.4	10.5	–	–	–	275

DME – Dimethoxyethane; THF – Tetrahydrofuran; THP – Tetrahydropyrane; 3-Me-THF – 3-Methyl-tetrahydrofuran

Table 10. The equilibrium constant $K_{t,\ell}$, at 25 °C, and $\Delta H_{t,\ell}$ and $\Delta S_{t,\ell}$ for inter-conversion of tight to loose sodium polystyryl ion-pairs in a variety of ethereal solvents

Solvent	$K_{t,\ell}$ at 25 °C	$\Delta H_{t,\ell}$ in kcal/mol		$\Delta S_{t,\ell}$ in e.u.		Ref.
		kinetic	conductance	kinetic	conductance	
DME	0.13	− 5.5	− 5.1	− 22.5	− 21.7	304, 303
THF	$2.25 \cdot 10^{-3}$	− 6.5	− 5.9	− 34	− 31	289, 303, 304
3-Me-THF	$5.8 \cdot 10^{-4}$	− 5.1	5.3	− 32	− 32	289 b
THP	$1.3 \cdot 10^{-4}$	− 3.0	–	− 28	–	300
Dioxane	$< 10^{-5}$	–	–	–	–	–

DME – Dimethoxyethane; THF – Tetrahydrofuran; THP – Tetrahydropyrane; 3-Me-THF – 3-Methyl-tetrahydrofuran

$$(A^-, Cat^+)_{tight} \overset{K_{t,\ell}}{\rightleftharpoons} (A^-, //, Cat^+)_{loose} \overset{K^*_{diss}}{\rightleftharpoons} A^- + Cat^+ .$$

$K_{t,\ell}$ denotes again the equilibrium constant of inter-conversion, while K^*_{diss} is the dissociation constant of loose ion-pairs. The latter is related to the overall dissociation constant K_{diss} by

$$K_{diss} = K^*_{diss}K_{t,\ell}/(1 + K_{t,\ell}) .$$

At very low temperatures $K_{t,\ell} \gg 1$; hence, $K_{diss} \approx K^*_{diss}$ and the slope of a van't Hoff line approaches the $- \Delta H^*_{diss}/R$ value. At the highest temperature, when $K_{t,\ell} \ll 1$, the $K_{diss} \approx K^*_{diss} \cdot K_{t,\ell}$ and the slope of the van't Hoff plot tends to the $- (\Delta H^*_{diss} + \Delta H_{t,\ell})/R$ value. The four thermodynamic parameters, $\Delta H_{t,\ell}$, $\Delta S_{t,\ell}$, ΔH^*_{diss} and ΔS^*_{diss} could be derived by computer fitting the function

$$\ell n K_{diss} = \exp[-(\Delta H^*_{diss} + \Delta H_{t,\ell})/RT]\exp[(\Delta S^*_{diss} + \Delta S_{t,\ell})/R]/ \\ \{1 + \exp(-\Delta H_{t,\ell} + T\Delta S_{t,\ell})/RT\}$$

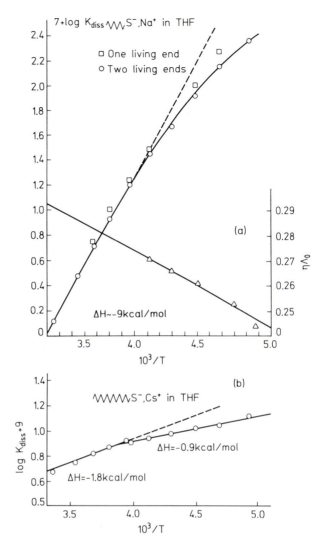

Fig. 46. a Van't Hoff plot of dissociation constant, K_{diss}, of sodium polystyryl pairs in THF. **b** Van't Hoff plot of dissociation constant, K_{diss}, of cesium polystyryl pairs in THF

to an appropriate, experimentally obtained van't Hoff plot. The constancy of these four parameters within the investigated temperature range is tacitly assumed. By this approach, Schulz and his cowerkers[303, 304] determined the $\Delta H_{t, \ell}$'s and $\Delta S_{t, \ell}$'s for the systems sodium polystyrene-DME and sodium polystyrene-THF. Their values, included in Table 10, satisfactorily agree with those derived from the kinetic studies. The heats of dissociation of the loose sodium polystyrene ion-pairs were found to be ~ 0 for the THF and 3-Me-THF systems and ~ -1 kcal/mol for the DME system, the respective entropies of dissociation being in the range -20 to -25 eu.

The following conclusions transpire from those combined kinetic and conductance studies:

(a) The propagation constants k_-, k_t and k_ℓ, as well as their activation parameters, are only slightly affected by the nature of an ethereal solvent. Significantly, k_t's are

$\sim 10^4$ times smaller than k_ℓ's and E_t's are by ~ 4 kcal/mol greater than the respective E_ℓ's, but A_t's and A_ℓ's are of a comparable magnitude.

(b) In contrast, $K_{t,\ell}$ is strongly affected by the nature of the solvent. Its value decreases from ~ 0.1 in the DME system to less than 10^{-5} in dioxane. The great variations of $K_{t,\ell}$ values are responsible for the large span of the observed $k_\pm$ values.

The virtual lack of solvent effect upon k_- and k_ℓ is readily rationalized. The interaction of ethereal solvents with free carbanions is weak, and therefore solvent molecules virtually do not interfere with the addition of monomers to those kind of growing centers. Similarly, monomer addition to growing loose ion-pairs takes place with only slight distortion of their structure as shown by the diagram:

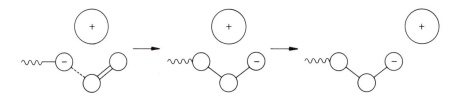

On the other hand, it is difficult to account for the virtual lack of dependence of k_t's on the solvent's nature, especially in the light of its very large influence upon $K_{t,\ell}$. Apparently, the degree of solvation of the transition state of the addition to tight pairs does not differ much from that of the initial state; in other words, the transition state resembles the initial state. Compensating effects could also play some role in the propagation of tight pairs. The incoming molecule of monomer has to displace a solvent molecule externally complexed to the pair. The stronger the solvent-pair interaction, the more difficult is its displacement; but this retarding effect is compensated by the more facile partial dissociation of the pair that is required in the addition process.

Let us close this section with brief discussion of some problems associated with determination of $k_\pm$ values. The experimental technique adopted by Szwarc and his coworkers in their kinetic studies of anionic polymerization greatly differs from that employed by Schulz' group. The former team used the spectrophotometric method to determine the concentration of living polymers and to follow their propagation. They worked, therefore, at a relatively low concentration of monomers. Schulz' team determined the concentration of living polymers by weighing the precipitated polymer and determining its number average molecular weight and therefore a relatively high concentration of monomer was maintained in their runs. Many of their experiments utilized a greatly improved flow technique, previously adapted for that purpose by Geacintov et al.[263]. Its most refined design was described recently by Löhr, Schmitt, and Schulz[366].

In spite of all the differences in experimentation, both groups came to the same conclusion on all of the investigated subjects. Many of their numerical findings show a remarkable quantitative agreement. For example, the same value of 25 $M^{-1/2}s^{-1}$ was reported by both groups for the $k_- K_{diss}^{1/2}$ of sodium polystyrene in THF, and a similar degree of agreement is shown by the $k_- K_{diss}^{1/2}$ pertaining to the sodium polystyryl-THP system investigated over a wide range of temperatures (see p. 97). The k_p values for the cesium polystyryl-THF system reported by both teams are virtually identical as shown by Fig. 47, and the $k_\pm$ values for sodium polystyrene in DME agree within 20%[282, 303, 306], etc. Whenever a numerical discrepancy occurred, it rarely exceeded a factor of 2.

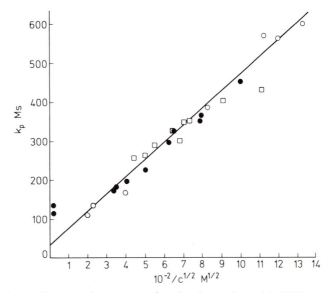

Fig. 47. Plot of the observed overall propagation constant, k_p, of cesium polystyryl in THF vs. $1/C_L^{1/2}$ (C_L concentration of living polymers). $\square$ – results of Löhr and Schulz[290]; ● – results of Bhattacharyya et al.[123]; $\bigcirc$ – results of Shimomura et al.[264]. Note their self-consistency

An instructive example of disagreement is provided by the determination of $k_\pm$ for the systems sodium polystyrene and cesium polystyrene in THF. These constants were obtained by extrapolation of a line giving k_p as a function of $1/C_L^{1/2}$ to $1/C_L^{1/2} = 0$. Such a line, involving the data-points obtained by both groups for the system cesium polysty-rene-THF[264 b, 123, 290], is shown in Fig. 47 and a similar plot was published by Tölle et al.[123, 321] for the data-points pertaining to the system sodium salt in THF. Their inspec-tion shows a good agreement between the observed values of k_p's; nevertheless, the $k_\pm$ values derived from such plots are greatly apart. While Schulz et al. claimed $k_\pm$ to be ~ 200 M^{-1}s^{-1} for the sodium salt and ~ 120 M^{-1}s^{-1} for the cesium salt, Szwarc' team reported the values of ~ 80 M^{-1}s^{-1} and ~ 25 M^{-1}s^{-1}, respectively, for the same con-stants.

The observation of Bhattacharyya et al. reported in their early paper (Ref. 264 b, p. 622) may explain the above discrepancy. They noted that k_p goes through a minimum and then *increases* as the concentration of electrolytes becomes high. Similar observations were made during later work and, in fact, such a trend is seen in Fig. 47; the k_p's obtained at the highest concentrations of cesium polystyrene seem to be too large. Careful exami-nation of Schulz' data reveals a similar trend in at least two cases, in the experiments depicted in Fig. 4 of a paper by Alvarino et al.[306], and again in those shown in Fig. 3 of a paper by Löhr and Schulz[310]. The latter figure includes also the k_p's plotted *vs.* $1/C_L^{1/2}$ for the experiments performed in the presence of NaBPh$_4$. Such plots, frequently seen in the literature*, are somewhat misleading. The pertinent k_p'a *are not* determined by $1/C_L^{1/2}$

---

* Regretfully, such a plot is given also in one of the papers of this writer[326]. Significantly, extrapola-tion of k_p *vs.* $1/C_L^{1/2}$ for the polymerization of sodium polystyrene in THP led to $k_\pm < 5$ M^{-1}s^{-1} [299], whereas a value of ~ 12 M^{-1}s^{-1} was obtained in the presence of NaBPH$_4$ (see Fig. 3 of Ref. 326)

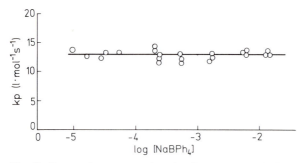

Fig. 48. Propagation constant, $k_\pm$, of sodium poly-α-methylstyrene ion-pairs in THP vs. the concentration of added sodium tetraphenylboride

but by $1/[Na^+]$, the relatively high concentration of sodium ions arises from the addition of NaBPH$_4$. Hence, the respective points should cluster close to the y-axis of the plot, revealing an increase of k_p at high concentrations of electrolytes.

In conclusion, the effect of high concentration of electrolytes upon the rate of anionic propagation deserves exploration. If it is real, it affects extrapolation and may lead to too high values of $k_\pm$. The only evidence to the contrary is provided by Dainton and Ivin[268], who determined k_p of sodium polystyrene in THP at constant concentration of living polymers but variable concentration of sodium tetraphenyl-boride. Their results are shown in the self-explanatory Fig. 48. It might be beneficial to have similar data for other solvent systems.

Although the preceding discussion of tight and loose ion-pairs was concerned with sodium polystyrene only, there is evidence for their participation in polymerization of other polystyrene salts. For example, Shimomura et al.[123] concluded that loose ion-pairs contribute to polymerization of cesium polystyrene in DME, and a stronger evidence for their participation in that reaction is furnished by the recent work of Bunge et al.[308]. Interestingly, the k_p's reported by Bunge et al. are higher than those found by Shimomura et al. This may be another manifestation of the effect of high electrolyte concentration upon the rate of propagation. The former workers perform their experiments in the presence of highly soluble CsBPh$_3$CN salt, i.e. at relatively high concentrations of electrolytes, whereas salts were not added to solutions of cesium polystyrene investigated by the latter workers. The activation energies reported by both teams, deduced from the linear Arrhenius plots, are closely similar, namely 4 kcal/mol and 4.5 kcal/mol, respectively.

IV.6. Rates of Dissociation of Ion-Pairs and Rates of Conversion of Tight into Loose Pairs

Association of oppositely charged ions is diffusion controlled. Theoretical treatment of Debye[334], that takes into account the Coulombic attraction between the ions, leads to the following expression for the bimolecular association constant, k_a,

$$k_a = 4\pi N_A 10^{-3}(\mathscr{D}_+ + \mathscr{D}_-)(r_+ + r_-)\delta/(\exp\delta - 1) \text{ in } M^{-1}s^{-1} .$$

Here, $\mathscr{D}$'s and r's denote the diffusion constants and radia of the interacting ions, and δ is the factor accounting for their mutual attraction, viz.

$$\delta = z_+ \cdot z_- e^2/DkT(r_+ + r_-) .$$

This factor amounts to about 10–20 for mono-valent ions with radia $r_+ + r_- \geqslant 5$ A when they associate in ethereal solvents of dielectric constant $D = 5$–8; i.e. their rates of encounters are by order of magnitude larger than of similar neutral particles. Using the Stokes-Einstein expression for the diffusion constants $\mathscr{D} = kT/6\pi\eta r$, one finds

$$k_a = 2 \cdot 10^{-3}(RT/3\eta)(2 + r_+/r_- + r_-/r_+)\delta/(\exp\delta - 1)$$

which for $r_+ \approx r_-$ is reduced to

$$k_a = 8 \cdot 10^{-3}(RT/3\eta)\delta/(\exp\delta - 1) \text{ in } M^{-1}s^{-1} .$$

Since the viscosity of THF at 25 °C is $4.5 \cdot 10^{-3}$ P, this approximate relation gives $k_a \sim 1.5 \cdot 10^{10}\delta/(\exp\delta - 1)$ for that solvent, while for THP, viscosity $7.6 \cdot 10^{-3}$ P, it leads to $k_a \sim 0.8 \cdot 10^{10}\delta/(\exp\delta - 1)$, both in units of $M^{-1}s^{-1}$.

The association of bare, "non-solvated" ions undoubtedly proceeds without any activation energy and without increase of the entropy of activation. Therefore, the respective $k_a > 10^{11} M^{-1}s^{-1}$. Ions surrounded by tight solvation shells form loose, solvent separated ion-pairs on their encounters, and again it is unlikely that their formation requires activation energy or negativ entropy of activation, i.e., most probably the pertinent $k_a > 10^{11} M^{-1}s^{-1}$. Temperature dependence of k_a is determined by temperature dependence of viscosity given by a formal activation energy of 2–3 kcal/mol.

It is instructive to calculate the lifetime of free anions under typical conditions prevailing in anionic polymerization. For living polymers of concentration of $\sim 10^{-3}$ M and for $K_{diss} \sim 10^{-7}$ M, the concentration of free anions is $\sim 10^{-5}$ M and their lifetime $\sim 10^{-6}$ s for $k_a = 10^{11} M^{-1}s^{-1}$. Assuming a value of $10^5 M^{-1}s^{-1}$ for their propagation constant, and considering 1 M solution of monomer, one finds that only 1 out of 10 anions has a chance to add one molecule of monomer during its lifetime. This is a typical result for polymerization of sodium polystyrene in THF at 25 °C, although under those conditions $\sim 90\%$ of polymerized monomer is added to the free anions, and only $\sim 10\%$ to their ion-pairs. Similar calculations show that under the same conditions the lifetime of the free anions in THP is ~ 30 times longer than in THF, allowing them to add ~ 3 molecules of monomer during their lifetime.

Experimental studies of ions association became feasible with the advent of relaxation techniques, picosecond spectroscopy, and ESR spectroscopy. Pioneering work in this field was done by Eigen, e.g.[298], who studied the association of inorganic ions in aqueous solutions. Recent development of field-modulation technique[335] allows for detailed investigations of ions association in low polarity media. This technique requires very low concentrations of ions in the investigated systems – a condition met by solutions involving low polarity solvents. Its application to systems such as tetrabutyl ammonium picrate in benzene-chlorobenzene mixture, or its bromide in benzene-nitrobenzene mixture, led to $k_a \sim 3 \cdot 10^{11}$ M^{-1}s^{-1} and $0.7 \cdot 10^{11}$ M^{-1}s^{-1}, respectively, at ambient temperature[336].

An interesting, and important for polymer chemists approach to the problem of ions association was developed by Schulz and his co-workers. Participation of two or more species in anionic propagation, each growing with its own specific rate constant, results in a broader molecular weight distribution of the formed polymer than Poisson distribution. The degree of deviation, measured by $\overline{DP}_w/\overline{DP}_n - 1 = U_{obs}$, depends on the rate of inter-conversion of the growing species as well as on their individual propagation constants[262]. Provided that only 2 species participate in the reaction, e.g. ion-pairs and free ions, one finds

$$U_{obs} = U_{Poiss} + U_{diss}$$

where $U_{diss} = (k_-^2 K_{diss}/k_a k_p) C_{LP} (2 - x_p)/2 x_p$.

Here, C_{LP} is the concentration of living polymers, k_p the observed propagation constant, k_- the propagation constant of free ions, k_a their association constant, and x_p is the ratio (polymerised monomer/M$_0$. Since all the parameters appearing in this expression except k_a were determined, the value of k_a could be calculated from the experimentally derived U_{diss}. This relation was used in the earlier work of Schulz; but since two kinds of ion-pairs participate in the process, a modified treatment was needed. Such refined treatment was reported by Böhm[337] and led to complex expression given in Ref. 262. Its application yields the values $k_a = 1.5 \cdot 10^{10}$ M^{-1}s^{-1} and $k_d = 10^4$ s^{-1} for the association constant of Na$^+$ and polystyryl$^-$ ions in THP at 25 °C, and for the corresponding dissocia-tion constant of their ion-pairs. Extension of these studies to lower temperature, down to -40 °C, yielded activation energy of the association $E_a = 4.7$ kcal/mol. It seems that k_a is somewhat too low and E_a too high.

The rate constants of conversion of loose pairs into tight, $k_{\ell,t}$, and tight into loose, $k_t, k_{\ell,t}$, were determined by a similar approach. Addition of an excess of sodium tet-raphenyl boride suppressed the dissociation of ion-pairs into free ions, and the broaden-ing of molecular weight distribution of the resulting polymer had to be attributed to the simultaneous participation in the propagation of tight and loose ion-pairs. The results summarized in review articles[322, 305] led to the constants listed in Table 11.

The results derived from studies of molecular weight distribution should be compared with those obtained by the ESR technique. By virtue of its character, the ESR technique requires paramagnetic salts as the subjects of investigation. Much work on behavior and

Table 11. Interconversion constants of tight into loose pairs, $k_{t,\ell}$, and loose into tight pairs, $k_{\ell,t}$

Solvent	T °C	$k_{t,\ell}$ s	$10^{-4}k_{\ell,t}$ s	$10^4 K_{t,\ell}$
THP	1	2.3	1.2	2.0
THP	− 43	0.6	0.1	5.7
THF	45	130.0	11.5	11.0
THF	25	105.0	4.7	22.0
THF	− 40	82.0	0.18	460.0
DME	20	$> 10^3$	> 1	1,400.0

THP – Tetrahydropyrane; THF – Tetrahydrofuran; DME – Dimethoxyethane

structure of ion-pairs was performed with salts of aromatic radical-ions. In fact, the first positive evidence for the existence of ion-pairs as distinct and lasting species was obtained by Atherton and Weissman[338] from their ESR studies of lines characterizing the sodium naphthalenide pairs. The line splitting was correctly interpreted as evidence of coupling of the odd electron to the nucleus of the cation. This observation not only identified the sodium naphthalenide pair as an entity but demonstrated that its lifetime is relatively long, at least 10^{-5} s.

Extensive studies of Hirota[339] demonstrated the existence of two kinds of lithium anthracenide pairs. Their ESR spectrum is composed of two superimposed spectra, one with large lithium coupling constant, the other with no additional splitting but distinct from the spectrum of the free anthracenide anion. Similar observations were reported in his earlier papers[341], and quantitative data on equilibrium and rates of inter-conversion were obtained from the temperature dependence of the spectra. Redoch[340] observed two superimposed spectra of sodium naphthalenide in THF, one showing splitting of the lines caused by cation (a_{Na} = 1.036 G), while the other did not. Since their relative intensities were affected by dilution, the equilibrium has to involve free ions and their pairs. Equilibrium involving three species, free ions and two kinds of ion-pairs was observed in the sodium-triphenylene system[342]. A comprehensive review of this subject was published by Sharp and Symons[343].

Conversion of tight ion-pairs into loose ones caused by the addition of a powerful solvating agent, tetraglyme, was observed by Höfelmann et al.[344]. The reaction is represented by the equation,

Sodium naphthalenide + glyme $\rightleftharpoons$ sodium, glyme, naphthalenide ,

and its equilibrium constant and forward and backward rate constants were obtained from examination of the ESR spectra, viz. 200 M^{-1}, $\sim 10^7 M^{-1}s^{-1}$, and $\sim 10^5 s^{-1}$, respectively.

IV.7. Effect of Pressure on Rates of Ionic Polymerization

Pressure affects rates of bimolecular reactions because the volume occupied by the reacting molecules is greater before their encounter than in their transition state. In most cases this effect is small, usually 2–3 ml/mol. Pressure affects also chemical equilibria – the state of lower volume is favored at higher pressure.

Anionic polymerization involves free ions and various kinds of ion-pairs as the reacting species; therefore, its rate depends on the degree of dissociation of ion-pairs into free ions, as well as on the degree of conversion of tight into loose ion-pairs. Free ions propagate polymerization of living polystyrene more rapidly than loose pairs, and the latter are substantially more reactive than the tight ion-pairs. Hence, the rate of living polystyrene propagation may be greatly affected by shifts in the pertinent equilibria caused by pressure.

What is the effect of pressure on those equilibria? Interaction of free ions and ion-pairs with polar solvent is caused by the electric field generated by their charge or dipole. Such an interaction compresses the solvent surrounding the ionic species, an effect known as electrostriction. Its degree increases with increasing strength of the field, being the greatest for free ions, smaller for loose ion-pairs, and the smallest for the tight pairs. Therefore, an increase of pressure should strongly enhance dissociation of ion-pairs into free ions and, to a lesser extent, the interconversion of tight ion-pairs into loose ones. Both effects are expected to speed up anionic polymerization of living polystyrene.

Le Noble[345] and independently Claesson, Lundgren, and Szwarc[346] investigated the effect of pressure on the equilibria of interconversion of tight into loose ion-pairs. Studies of Le Noble covered pressure changes up to 1,000 atm and were limited to the system lithium fluorenyl-THF. Claesson et al. extended their work up to 5,000 atm, investigating lithium and sodium fluorenyl in THF, as well as sodium fluorenyl in DME and lithium fluorenyl in THP and Me-THF. The primary results of both groups pertaining to the lithium fluorenyl-THF system are in good agreement. The following conclusions were derived from their work: Conversion of tight ion-pairs of fluorenyl salts into loose ones is accompanied by a large decrease of volume of the system, namely, -15.6 ml/mol and -24.4 ml/mol for the lithium and sodium salts in THF, -21 ml/mol for sodium fluorenyl in DME, and -11 ml/mol and -23 ml/mol for lithium fluorenyl in THP and 2-Me-THF, respectively. Moreover, these findings allowed Claesson et al. to estimate the magnitude of the "pressure" compressing the solvent around the ion-pairs at $\sim 2,000$–$2,500$ atm.

The effect of pressure on the rate of polymerization of sodium polystyryl in THP and cesium polystyryl in DME was investigated by Bunge, Höcker, and Schulz[347]. Polymerization was carried out in the presence of homo-ionic salts, viz. $NaBPH_4$ and $CsBPh_3CN$, respectively. Since an excess of those salts was added, it was assumed that the observed propagation constant is equal to the overall propagation constant of ion-pairs, $k_\pm$, given by

$$k_\pm = (k_t + k_\ell K_{t,\ell})/(1 + K_{t,\ell}) \ .$$

Using the symbolism of the preceding section, one finds the $k_\pm$ to be a function of E_t, A_t, E_ℓ, A_ℓ, $\Delta H_{t,\ell}$, $\Delta S_{t,\ell}$ as well as $\Delta V_t^\ddagger$, $\Delta V_\ell^\ddagger$ and $\Delta V_{t,\ell}$. The last three parameters denote activation volumes of propagation of tight and loose ion-pairs, and the volume change arising from conversion of tight into loose ion-pairs. The experimental data were presented in the form of plots of log $k_\pm$ vs. pressure at constant temperature and plots of log $k_\pm$ vs. 1/T at constant pressure. The former plots were linear, while the latter were curved.

Two explanations could account for the linearity of plots of log $k_\pm$ vs. pressure. Either $\Delta V_t^\ddagger \approx \Delta V_\ell^\ddagger + \Delta V_{t,\ell}$ or $k_t \ll k_\ell K_{t,\ell}$ and $K_{t,\ell} \ll 1$. Bunge et al. preferred the first alternative and calculated $\Delta V_t^\ddagger$ and $\Delta V_\ell^\ddagger + \Delta V_{t,\ell}$ as -22.2 ml/mol and -27.4 ml/mol, for the system sodium polystyryl-THP, and -27.9 ml/mol and -25.6 ml/mol*, respectively, for the cesium polystyryl-DME system. Implicit in these calculations is the assumption that the parameters E_t, A_t, E_ℓ, A_ℓ, $\Delta H_{t,\ell}$, and $\Delta S_{t,\ell}$ are independent of pressure, being given by the results of previous studies of Schulz' group (see Tables 9, 11). The claimed values for $-\Delta V_t^\ddagger$ seem to be too large. They imply the transition state of propagation by tight ion-pairs to be similar in its structure to the loose pairs.

---

* The results pertaining to cesium polystyrene imply a contraction of volume in the transition state of tight pairs to be greater than in their conversion into loose pairs, i.e. $|\Delta V_t^\ddagger| > |\Delta V_{t,\ell}|$, or an *expansion* of the volume in the transition state of the loose pair, i.e. $\Delta V_\ell^\ddagger > 0$

This writer favors the second explanation. It is possible that the k_t values claimed by Schulz are somewhat too large and $k_\pm = (k_t + k_\ell K_{t,\ell})/(1 + K_{t,\ell}) \approx k_\ell K_{t,\ell}$, especially at higher pressures. In this interpretation the pressure dependence of $k_\pm$ virtually is governed only by $\Delta V_{t,\ell}$ leading to their reasonable values of -20–25 ml/mol. In conclusion, whatever the explanation, a large effect of pressure on the rate of anionic polymerization is revealed by the work of Bunge et al.

IV.8. Role of Triple-Ions in Ionic Polymerization

The concept of triple-ions arose from the conductance studies of Fuoss and Krauss[297] who observed an increase of equivalent conductance of electrolyte solutions at higher salt concentrations. They interpreted these findings as evidence of association of the free ions with their pairs, viz.

$$A^- + B^+, A^- \rightleftharpoons A^-, B^+, A^-, \quad K_{t-},$$

and

$$B^+ + A^-, B^+ \rightleftharpoons B^+, A^-, B^+, \quad K_{t+}.$$

The two equilibria shown above may be combined into one when $K_{t-} = K_{t+}$, viz.

$$3 A^-, B^+ \rightleftharpoons A^-, B^+, A^- + B^+, A^-, B^+,$$

and then the fraction of triple-ions $([A^-, B^+, A^-] + [B^+, A^-, B^+])/[A^-, B^+]$ increases proportionally with the square-root of ion-pairs concentration.

The concentration of free ions virtually is unaffected by the formation of triple-ions if $K_{t-} = K_{t+}$, provided that the proportion of the latter ions is low. However, when $K_{t-} > K_{t+}$ or $K_{t+} > K_{t-}$, the formation of triple-ions affects the concentration of free ions. For example, the ratio $[A^-]/[B^+] > 1$ when $K_{t+} > K_{t-}$, a result required by the condition of electric neutrality of the solution.

Let us consider an extreme case, namely $K_{t+} \approx 0$ and $K_{t-} \gg K_{t+}$. In such a system the equilibrium,

$$2 A^-, B^+ \rightleftharpoons A^-, B^+, A^- + B^+, \quad K_{T+},$$

with $[A^-, B^+, A^-] \approx [B^+]$, governs the formation of the triple A^-, B^+, A^- ions; and at a sufficiently high concentration of ion-pairs the inequality

$$K_{T+} \gg K_{diss}/[A^-, B^+]$$

becomes valid. Hence, the generation of triple-ions yields more charged species than a direct dissociation of ion-pairs. Indeed, $[B^+]$ is larger than expected for a simple ion-pair dissociation and it is proportional to $[A^-, B^+]$. Its buffering effect depresses the dissociation of ion-pairs, lowers $[A^-]$, and makes its concentration independent of $[A^-, B^+]$, i.e. $[A^-] = K_{diss}/K_{t+}^{1/2}$.

The rate of anionic polymerization proceeding under such conditions is slower than expected for a mechnism ivolving the free ions and their pairs only. Its pertinent *first* order constant, $k_u = -d\ell n[\text{monomer}]/dt$, is independent of living polymers concentration, provided that the free A^- ions (the growing polymeric ions) are the only significant contributors to the propagation and the fraction of living polymers present in any form but ion-pairs is low. In fact, for such a polymerization, $k_u = k_-K_{diss}/K_{t+}^{1/2}$, and the conductance of such solutions of living polymers is high due to large concentration of the mobile B^+ ions.

In the reverse case when $K_{t-} \approx 0$ and $K_{t+} \gg K_{t-}$, the concentration of the B^+, A^-, B^+ ions is determined by the equilibrium

$$2A^-, B^+ \rightleftharpoons B^+, A^-, B^+ + A^-, \qquad K_{T-}$$

with $[B^+, A^-, B^+] \approx [A^-]$. In such systems the concentration of A^- ions is higher than anticipated from the simple dissociation of ion-pairs, provided that $K_{T-} \gg K_{diss}/[A^-, B^+]$. Propagation of anionic polymerization is faster under those conditions than predicted by a mechanism involving only the free ions and their pairs, and the *bimolecular* propagation constant, $k_p = k_-K_T^{1/2}$, is then independent of living polymer concentration. The conductance of such solutions of living polymers is low and their equivalent conductance Λ is independent of C_L because the mobile B^+ ions are replaced by the sluggish $\sim\!\!\!\sim A^-, (B^+)_2$ triple-ions.

A few words about reactivities of triple-ions are in place. The A^-, B^+, A^- triple-ions are more reactive than the pair A^-, B^+ but substantially less reactive than the free ion, A^-. Unfortunately, nothing is known about reactivities of B^+, A^-, B^+ triple-ions; most likely they are less reactive than the pairs.

In an interesting paper, Sigwalt et al.[352] suggested that a significant proportion of positively charged triple-ions is formed in the system sodium-poly-2-vinyl pyridine (pVPy, Na$^+$) in THF solution. Conductance of these solutions is extremely low[353-355], e.g., the respective K_{diss} determined at 25° is about 100 times lower than that of sodium polystyryl, viz. $\sim 10^{-9}$ M compared with $\sim 10^{-7}$ M. Significantly, a higher degree of dissociation was observed for the salt of poly-4-vinyl pyridine[354]. It was proposed, therefore, that an intramolecular solvation by the pyridine moieties of the polymer determines the structure of pVPy$^-$, Na$^+$ ion-pairs as shown below:

an interaction sterically precluded for the 4-vinyl pyridine isomer. This interaction hinders the release of the free, mobile sodium ions and accounts for the low conductance of pVPy$^-$Na$^+$ solutions. The proposed penultimate unit effect disappears when styryl, instead of vinyl pyridine, segments precede the last 2-vinyl-pyridine$^-$ end-group. Indeed, the degree of dissociation of sodium salt of polystyrene terminated by a single 2-vinyl-pyridine$^-$ anion end-group is higher than of sodium poly-2-vinyl pyridine[353].

The low degree of dissociation of pVPy$^-$, Na$^+$ pairs favors the formation of triple-ions. Kinetics of their propagation, reported by Fisher and Szwarc[353] and verified by Sigwalt et al.[352], reveals some unusual features. The bimolecular propagation constants, k_p, although very large, are only slightly affected by the variation of concentration of living polymers. For example, the conventional plot of k_p vs. $1/C_L^{1/2}$, derived from the data obtained at 25 °C, has a slope of $\sim 3\,M^{-1/2}\,s^{-1}$ with an intercept of 2,100 $M^{-1}s^{-1}$, and the plots are nearly horizontal at lower temperatures. This peculiar behavior of pVPy$^-$, Na$^+$ may be explained by the formation of positive triple-ions since, as pointed out earlier, their presence leads to fast polymerization with a nearly constant k_p.

To shed more light on that problem, Sigwalt et al.[352] investigated the polymerization of cesium poly-2-vinyl pyridine endowed with only one growing end-group. This salt is more dissociated in THF than the sodium salt, contrary to the order found in other systems. Apparently the cesium salt has a conventional structure; the large size of the Cs$^+$ cation prevents its placement inside the cavity formed by the ultimate and penultimate pyridine moiety. The kinetics of propagation of the cesium salt also appears to be conventional. The results, interpreted by the simple mechanism involving the free ions and their pairs, leads to the following values of k_- and k_+:

T°C	15	0	− 20
$10^{-5}k_-$ M s	1.4	1.05	0.27
Cs$^+$; $10^{-3}k_\pm$ M s	1.5	1.1	0.4

A similar treatment of the data reported for the sodium salt[353] leads to

T°C	25	0	− 20	Ref.
$10^{-5}k_-$ M s	1 ± 0.3	0.14 ± 0.04	0.06 ± 0.02	353
	−	0.15	−	352
Na$^+$; $10^{-3}k_\pm$ M s	2.1	0.72	0.28	353

As correctly pointed out by Sigwalt, had the same mechanism been operating in both systems, the same k_- values should be derived from either set of data. This is not the case, implying that the mechanism operating in the sodium system is different from that governing the polymerization of the cesium salt, and the participation of triple-ions in the sodium system looks like an attractive suggestion.

Closer examination of all the data clouds this conclusion. The presence of triple-ions is significant provided that $K_{T^-} \gg K_{diss}/C_L$. The lower limit for K_{T^-} could be estimated by assuming $K_{T^-} = 4\,K_{diss}/C_L$. For a relatively large $C_L = 10^{-4}$ M, the following values are calculated from data given in Ref. 353 in conjunction with the k_- values of Sigwalt:

T °C	25	0	− 20
$10^6 K_{T^-}$	36	56	80
$k_- K_{T^-}^{-1/2}$ M s[a]	840	750	810
k_p M s	2,300	~ 800	~ 300
$k_\pm$ M s[b]	2,100 (?)	~ 700 (?)	~ 300 (?)

[a] This is the contribution of free anions to the observed propagation constant, k_p
[b] They do not represent $k_\pm$ if indeed a triple-ion mechanism operates

The reliability of the $k_\pm$ values reported in Ref. 353 could be questioned, although Sigwalt reported a similar value of 700 $M^{-1}s^{-1}$ for $k_\pm$ at 0 °C based on his results derived from the experiments performed in the presence of NaBPh$_4$. The contribution of poly-2-vinyl pyridile ions calculated on the basis of triple-ion participation is, therefore, too large. Apparently, more work is needed to clarify the problem.

Intramolecular formation of negative triple-ions in THF was observed in the system cesium polystyrene endowed with two propagating end-groups[264 b, 295]. Those polymers propagate more slowly than the polymers possessing only one propagating end-group and the conductance of their solutions is lower than that of the monofunctional ones. For example, slopes of the plots giving k_p as a function of $1/C_L^{1/2}$ are 1.4 $M^{-1/2}s^{-1}$ for the di-anions of cesium polystyrene of DP ~ 25, but 3.0 $M^{-1/2}s^{-1}$ for the salt of mono-anions; both constants refer to 25 °C. The conductance data yield $1/\Lambda_\infty^2 K_{diss} = 9.0 \cdot 10^{-3}$ for the former polymers while a substantially larger value of $53 \cdot 10^{-3}$ is found for the latter. Hence, the apparent dissociation constant of the di-anionic cesium polystyrene is calculated as $0.00465 \cdot 10^{-7}$ M from the conductance data whereas calculation based on the kinetic data gives the apparent $K_{diss} = 0.165 \cdot 10^{-7}$ M. The correct value of K_{diss}, consistently derived either from the kinetic or from the conductometric study of mono-anions of cesium polystyrene, is $0.02–0.03 \cdot 10^{-7}$ M.

This apparent anomaly was accounted for by the following mechanism:

$$Cs^+, \bar{S}\!\!\sim\!\!\bar{S}, Cs^+ \rightleftharpoons Cs^+, \bar{S}\!\!\sim\!\!\bar{S} + Cs^+ , \qquad\qquad K_{diss} ,$$

$$\bar{S}\!\!\sim\!\!\bar{S}, Cs^+ \rightleftharpoons (\bar{S}, Cs^+, \bar{S}) , \qquad\qquad K_c$$

where $\sim\!\!\bar{S}$ denotes the anionic end-group of living polystyrene. The cyclization consumes the $\sim\!\!\bar{S}$ ions, and their replenishment by the dissociation of $Cs^+, \bar{S}\!\!\sim\!\!\bar{S}, Cs^+$ results in a too high concentration of Cs^+ ions – the main contributors to conductance. Consequently, the polymerization is too slow because the most reactive free $\sim\!\!\bar{S}$ ions are depleted, while the conductance is too high due to a large concentration of the mobile Cs^+ cations. The available data allow the calculation of the cyclization constant, $K_c = 5.5$, and the propagation constant of triple-ions as 2,200 $M^{-1}s^{-1}$; i.e., the triple-ions popagate ~ 100 times faster than the $\sim\!\!\bar{S}, Cs^+$ ion-pairs but about 30 times slower than the free $\sim\!\!\bar{S}$ ions.

The proposed mechanism predicts the reactivity of di-anionic cesium polystyrene to increase on increasing length of its chain. This, indeed, was observed[264 b].

A direct evidence for the existence of triple-ions has been provided by the ESR studies of THF solutions of sodium salts of duroquinone radical anions in the presence of sodium tetraphenylboride[325, 351]. Under those conditions each line of the spectrum arising from proton-splitting of the signal becomes further split into seven lines caused by the interaction of the odd electron with the nuclei of two sodium cations (total spin 3). The sharpness of these lines implies a lifetime of triple-ions longer than 10^{-5} s.

IV.9. Propagation of Living Polymers Associated with Bivalent Cations

Two distinct classes of living polymers salts with bivalent cations should be distinguished. Living polymers endowed with a single active end-group associate with bivalent cations into linear, neutral aggregates capable of dissociating into two electrically charged fragments that conduct current, viz.

$$\text{\Large$\sim\!\!\!\sim$} \overline{M}, Cat^{2+}, \overline{M} \text{\Large$\sim\!\!\!\sim$} \;\rightleftharpoons\; \text{\Large$\sim\!\!\!\sim$} \overline{M}, Cat^{2+} + \text{\Large$\sim\!\!\!\sim$} \overline{M}.$$

On the other hand, those possessing two active end-groups, on both ends, associate with bivalent cations into cyclic aggregates. Their rings may open yielding a linear form, viz.

$$\overline{M}, Cat^{2+}, \overline{M}_\rceil \;\leftrightarrows\; Cat^{2+}, \overline{M} \text{\Large$\sim\!\!\!\sim$} \overline{M},$$

that does not contribute to the conductance. The equilibrium between the cyclic and the open, linear forms is unaffected by variation of living polymers concentration. However, it shifts to the right when the length of the connected chain increases.

Still larger, cyclic aggregates could be formed, e.g.,

$$\overline{M}, Cat^{2+}, \overline{M} \text{\Large$\sim\!\!\!\sim$} \overline{M}, Cat^{2+}, \overline{M}_\rceil \;,\; \overline{M}, Cat^{2+}, \overline{M} \text{\Large$\sim\!\!\!\sim$} \overline{M}, Cat^{2+}, \overline{M} \text{\Large$\sim\!\!\!\sim$} \overline{M}, Cat^{2+}, \overline{M}_\rceil,$$

etc. The abundance of larger cyclic forms, involving two or more bivalent cations, decreases on dilution of the salt's solution.

The behavior of barium polystyryl endowed with a single carbanionic endgroup, $\text{\Large$\sim\!\!\!\sim$} \overline{S}$, $Ba^{2+}, \overline{S} \text{\Large$\sim\!\!\!\sim$}$, was extensively investigated[56]. Such a salt was prepared by reacting dibenzyl barium dissolved in THF with a small amount of α-methyl styrene and thereafter with larger amounts of styrene. This indirect procedure was necessary because benzyl salts slowly initiate styrene polymerization while its propagation is fast. This could lead to incomplete conversion of benzyl anions into polystyryl anions. On the other hand, the propagation is slow in the α-methyl styrene case, and moreover, the equilibrium concentration of this monomer with its living polymer is relatively high. As a result, the conversion of benzyl anions into poly-α-methyl styrene anions is quantitative, and since the latter rapidly initiate polymerization of styrene, the desired polymer is readily formed.

The propagation ensuing on addition of fresh styrene to the barium salt was investigated by the conventional spectrophotometric technique. As expected, plots of ℓn[styrene] versus time were linear but, surprisingly, their slopes determining the pseudo-first order propagation constant were independent of the salt concentration, at least within

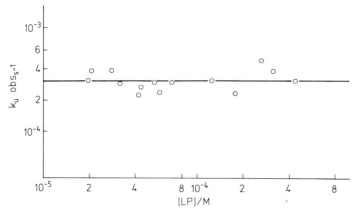

Fig. 49. Plot of the pseudo-first order propagation constant $k_u = -d\ell n[M]/dt$ for polymerization of barium salt of mono-ended polystyrene, $Ba^{2+} (\overline{S}\rightsquigarrow)_2$, vs. its concentration in THF. $T = 20\,°C$

the $10^{-5}-10^{-3}$ M range. This unexpected behavior, illustrated by Fig. 49, was rationalized by postulating two modes of ions formation. The conventional unimolecular association,

$$\rightsquigarrow \overline{S}, Ba^{2+}, \overline{S} \rightsquigarrow \rightleftharpoons \rightsquigarrow \overline{S}, Ba^{2+} + \rightsquigarrow \overline{S} , \qquad\qquad K_1 ,$$

yields the free polystyrene anions and the formal cations, $\rightsquigarrow \overline{S}, Ba^{2+}$, while the bimolecular reaction,

$$2(\rightsquigarrow \overline{S}, Ba^{2+}, \overline{S} \rightsquigarrow) \rightleftharpoons \rightsquigarrow \overline{S}, Ba^{2+} + (\rightsquigarrow \overline{S})_3 Ba^{2+} , \qquad\qquad K_2 ,$$

again produces the formal cations, $\rightsquigarrow \overline{S}, Ba^{2+}$; but instead of $\rightsquigarrow \overline{S}$ anions, the formal $(\rightsquigarrow \overline{S})_3 Ba^{2+}$ anions are formed. Provided that substantially more ions are formed by the second mode than by the direct dissociation, i.e.

$$K_2[(\rightsquigarrow \overline{S})_2 Ba^{2+}] \gg K_1 ,$$

and that the mole fraction of all the ions is small, the equilibrium concentration of the free $\rightsquigarrow \overline{S}$ ions is given by,

$$[\rightsquigarrow \overline{S}] = K_1/K_2^{1/2} ,$$

i.e. it is independent of the concentration of the salt. This relation reflects the buffering effect of the bimolecular mode of ion formation on the unimolecular dissociation. It is assumed, furthermore, that only the free $\rightsquigarrow \overline{S}$ ions are responsible for the propagation, the contribution of the other species, $(\rightsquigarrow \overline{S}, Ba^{2+})$, $(\rightsquigarrow \overline{S})_2 Ba^{2+}$, and $(\rightsquigarrow \overline{S})_3 Ba^{2+}$, being insignificant. This assumption was verified by investigating the effect of barium tetraphenyl-boride on the rate of propagation[57]. Denoting by k_p the bimolecular rate constant of styrene addition to the free $\rightsquigarrow \overline{S}$ ions, one derives

$$k_u = -d\ell n[\text{styrene}]/dt = k_p K_1/K_2^{1/2} \, ,$$

i.e., its value is constant, independent of the salt concentration, as required by the observations.

Conductance studies led to the determination of K_1 and K_2, the results justified the inequality $K_2[(\sim \overline{S})_2 Ba^{2+}] \gg K_1$ and confirmed the idea of two modes of ions formation.

The recently investigated propagation by strontium polystyrene[58] proceeds like that of the barium salt.

The barium or strontium salts of living polystyrene oligomers endowed with two carbanionic groups were prepared by Francois[59]. Styrene dissolved in tetrahydrofuran or tetrahydropyrane was reacted with the colloidal metaf prepared by the procedure described earlier[60]. The highly dispersed metal is extremely reactive, e.g. its wetting by ethers is strongly exothermic. Therefore, precautions were needed to prevent the hydride transfer to barium from the oligomeric, presumably dimeric, dicarbanions. Such a transfer converts the dicarbanions into a mono-carbanionic salt. Indeed, the earlier attempts to prepare the dicarbanionic oligomers by this procedure[61] were frustrated by the described above undesirable reaction[56], although the authors did not realize it at that time.

The dicarbanionic nature of the ultimately produced oligomers was confirmed by their conversion into dicarboxylic acids. Moreover, their optical spectrum reveals splitting of the absorption band due to the chromophore-chromophore interaction (see, e.g. Ref. 465), and this phenomenon provides an additional argument for their dicarbanionic structure. Nevertheless, this writer is not convinced that the prepared solutions used for the kinetic studies of styrene propagation were rigorously free of the monocarbanionic oligomers.

The kinetic results[62] showed that plots of the $\ell n[\text{styrene}]$ vs. time, reproduced in Fig. 50, to be sigmoidal, indicating a strong acceleration of the reaction with the increasing conversion. Such results were rationalized by assuming that only the open form of the cyclic salt is responsible for the propagation

$$\lceil \overline{S}, Ba^{2+}, \overline{S} \rceil \rightleftharpoons Ba^{2+}, \overline{S} \sim \overline{S}, \quad K_{op} \, .$$

Since the above equilibrium shifts to the right as the linking chain becomes longer, the observed acceleration is justified.

Determination of the $\overline{M}_w/\overline{M}_n$ ratio corroborates this suggestion. Since the larger rings propagate faster than the smaller ones, those polymers that started their growth earlier would grow faster than the other which started their propagation later. The molecular weight distribution becomes broader therefore than otherwise expected. The $\overline{M}_w/\overline{M}_n$ ratio was indeed substantially larger than 1.0.

Determination of the pseudo-first order propagation constants, i.e. $-d\ell n[\text{styrene}]/dt$, required the evaluation of tangents of the sigmoid curves – a difficult and not too precise task. Their values, according to the authors, seem to be independent of the salt concentration, provided that the results refer to rings of the same size, again a requirement difficult to fulfill in view of the broad molecular weight distribution.

Two mechanisms were proposed to account for these findings:
(a) The cyclic salt forms large aggregates

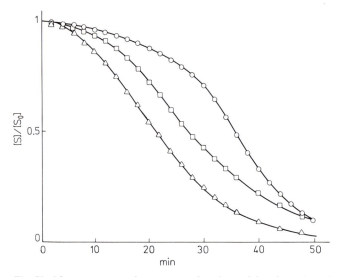

Fig. 50. Monomer conversion curves as functions of time for polymerization of styrene initiated by barium salt of polystyryl oligomers endowed with two active ends. Such a salt exists in a ring structure in equilibrium with minute fraction of an open form. Only the latter propagate. Solvent – THF. T = 20 °C. Concentration of active ends: $\bigcirc - 2.4 \cdot 10^{-4}$ M, [S] = $7.6 \cdot 10^{-2}$ M; $\square - 16.7 \cdot 10^{-4}$ M, [S] = $75 \cdot 10^{-2}$ M; $\triangle - 4.15 \cdot 10^{-4}$ M, [S] = $17.6 \cdot 10^{-2}$ M. Note the surprising and unexplained position of the curve representing $\square$ points between the other two curves. One should expect to find it below the other two curves

$$(\ulcorner \overline{S}, Ba^{2+}, \overline{S} \urcorner)_n \; \rightleftharpoons n \; (\ulcorner \overline{S}, Ba^{2+}, \overline{S} \urcorner) ; \quad K_n ,$$

remaining in equilibrium with a minute fraction of non-aggregated salt. The latter, presumably in their open form, propagates the reaction. This approach gives the required expression for large n,

$$- d\ell n[\text{styrene}]/dt = k_p K_{op} K_n^{1/n} .$$

However, it raises the question why a similar aggregation does not occur in the monocarbanionic systems. The conductance data obtained for the latter system, especially their dependence on concentration[56], disprove the hypothesis of aggregation of the monocarbanionic salt.

(b) The cyclic salt is in equilibrium with its open form, the only one that contributes to the propagation. Further more, an additional equilibrium is proposed:

$$Ba^{2+}, \overline{S} \wedge\!\wedge\!\wedge \overline{S} + \ulcorner \overline{S}, Ba^{2+}, \overline{S} \urcorner \rightleftharpoons Ba^{2+}, \overline{S} \wedge\!\wedge\!\wedge \overline{S}, Ba^{2+} \left| \begin{array}{c} \overline{S} \urcorner \\ \\ \overline{S} \end{array} \right. , \quad K_T ,$$

with an unconventional assumption, namely the above association equilibrium does not affect the equilibrium established between the open and closed form of the cyclic barium salt. On accepting this proposition, one finds

$$K_{op}[\overline{S}, Ba^{2+}, \overline{S}] = (1 + K_T [\overline{S}, Ba^{2+}, \overline{S}]) [Ba^{2+}, \overline{S} \sim \overline{S}] ,$$

and hence

$$[Ba^{2+}, \overline{S} \sim \overline{S}] = K_{op}[\overline{S}Ba^{2+}\overline{S}]/(1 + K_T [\overline{S}Ba^{2+}\overline{S}]) .$$

Provided that $K_T[\overline{S}Ba^{2+}\overline{S}] \gg 1$, the rate of polymerization, attributed to the free, unassociated $\sim \overline{S}$ ion, is independent of the salt concentration, i.e.

$$- d\ell n[styrene]/dt = k_{obs} = k_p \cdot K_{op} .$$

This treatment accounts for the experimental findings. Nevertheless, it is questionable because the association of the free $\sim \overline{S}$ ion with the cyclic form of the salt forms a new species, viz. a kind of triple ion. Since the concentration of the open, free $\sim \overline{S}$ ions is still given by $K_{op}[\overline{S}, Ba^{2+}, \overline{S}]$, the rate of propagation should be proportional to the concentration of the barium salt, whether a further association with the cyclic form does or does not take place, provided that the bulk of the salt is still present in the conventional cyclic form.

IV.10. Mixed Solvents and Solvating Agents

Anionic polymerization of lithium polystyrene in mixtures of THF and dioxane was investigated by Van Beylen et al.[327], and in the THF-benzene solutions by Worsfold and Bywater[328]. Physical properties of those solvents, such as density, viscosity and dielectric constants, were determined over the whole range of compositions and found to monotonically vary with the volume percent of THF.

Conductometric measurements were feasible down to ~ 50 volume percent of THF, and the dissociation constants of lithium ion-pairs were determined from those data. The log K_{diss}' were shown to be linear with reciprocal of the appropriate dielectric constant – a relation commonly reported[284] although not obvious. The solvating power of a medium is not determined uniquely by the *bulk* dielectric constant, D; the more sophisticated Kirkwood function $(D - 1)/(2D + 1)$[329] could provide a better measure of that property. However, since the Kirkwood function is nearly linear with $1/D$ in the investigated range of dielectric constants, the discrimination between them is not practical.

The observed propagation constants, k_p, were shown to vary linearly with $1/C_L^{1/2}$ within the range of compositions permitting the evaluation of K_{diss}. The slopes of those lines, in conjunction with the respective K_{diss}, allow the calculation of propagation constants of the free ions, k_-. For the reaction proceeding in 47% THF-dioxane at 25 °C, its value is $\sim 60,000$ $M^{-1}s^{-1}$, once again showing the negligible dependence of k_- on the nature of the ethereal solvent. However, in the THF-benzene mixtures, k_- decreases from $\sim 78,000$ $M^{-1}s^{-1}$ in pure THF to $\sim 40,000$ $M^{-1}s^{-1}$ for the propagation proceeding in a 50:50 mixture.

The propagation constant, $k_\pm$, of ion-pairs is strongfy affected by the composition of the solvent. In THF-benzene solutions its value decreases from 170 $M^{-1}s^{-1}$ to 0.6 $M^{-1}s^{-1}$ as the mole fraction of THF drops from 0.74 to 0.01. Plots of log $k_\pm$ vs. $(D - 1)/(2D + 1)$

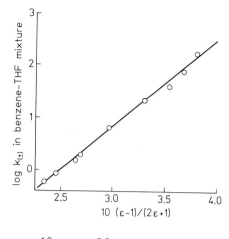

Fig. 51. Plot of log of the propagation constant, $k_\pm$, of lithium polystyryl ion-pairs in THF-benzene mixtures as a function of $(\varepsilon - 1)/(2\varepsilon + 1)$, ε – the dielectric constant of the mixture

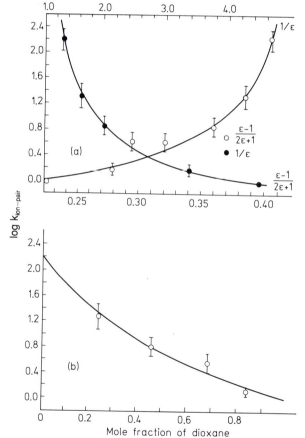

Fig. 52. a Plots of log of the propagation constant, $k_\pm$, of lithium polystyryl ion-pairs in THF-dioxane mixtures as a function of $1/\varepsilon$ – ●, and of $(\varepsilon - 1)/(2\varepsilon + 1)$ – ○. ε – the dielectric constant of the mixture. Note their concave shapes. **b** Plot of log of the propagation constant, $k_\pm$, of lithium polystyryl ion-pairs in THF-dioxane mixtures vs. mole fraction of dioxane. Note the concave shape of the curve

are linear, as shown in Fig. 51. The activation energy, $E_\pm$, increases from ~ 4 kcal/mol to ~ 10 kcal/mol as the mole fraction of THF decreases from 10% to 1%.

In THF-dioxane mixtures the plot of $k_\pm$ vs. mole fraction of dioxane is concave, as shown in Fig. 52, although the analogous plot of the dielectric constant is convex. Plotting of $k_\pm$ vs. $1/D$ or $(D-1)/(2D+1)$ results in even more concave curves. The concave shape of these plots is intriguing. Composition of the solvent around ions or ion-pairs differs from its average, bulk composition – the solvent around the pairs is expected to be richer in the component of greater solvating power. Therefore, one would expect a convex plot of $k_\pm$ vs. mole fraction of dioxane if THF is the better solvating agent. However, this unexpected behavior of those solutions is in line with Grunwald's findings[330], who reported a stronger interaction of small cations with dioxane than with water.

Addition of small amounts of powerful solvating agents to solutions of living polymers in relatively poor solvating media often exerts a large influence on the rates of their propagation. An example is provided by the work of Shinohara et al.[326, 331], who investigated the effects exerted by the addition of triglyme or tetraglyme to solutions of sodium polystyrene in tetrahydropyrane. As illustrated by Fig. 53, the observed propagation constants, k_p, are linear functions of $1/C_L^{1/2}$, at any constant concentrations of the glyme, but their intercepts and slopes are greatly affected by the nature and concentration of the glyme. For example, the intercept increases from 12 $M^{-1}s^{-1}$ determined in the absence of glyme to 3,500 $M^{-1}s^{-1}$ in the presence of $3 \cdot 10^{-3}$ M tetraglyme.

It seems that three kinds of species participate in the propagation: the "ordinary" ion-pairs existing in THP, the pairs associated and presumably separated by glyme, and the free polystyryl ions. Let the "ordinary" ion-pair be denoted by $\sim\!\!\sim S^-, Na^+$ and that associated with glyme by $\sim\!\!\sim S^-, G, Na^+$; G denoting the glyme. Two equilibria are established in addition to the conventional equilibrium between "ordinary" ion-pairs and Na^+ cations surrounded by THP molecules, namely,

$$\sim\!\!\sim S^-, Na^+ + G \rightleftharpoons \sim\!\!\sim S^-, G, Na^+ , \qquad\qquad K_G ,$$

$$Na^+ + G \text{ (or } 2G) \rightleftharpoons Na^+G \text{ (or } Na^+, 2G) , \qquad\qquad K_{G+} ,$$

as well as

$$\sim\!\!\sim S^-, Na^+ \rightleftharpoons \sim\!\!\sim S^- + Na^+ , \qquad\qquad K_{diss} .$$

Since, as verified *a posteriori*, 90% or more of living polymers are in the form of "ordinary" ion-pairs, provided that the concentration of glyme is not too high, the observed propagation constant, k_p, is given by

$$k_p = \{k_\pm + k_G K_G[G]\} + k_- K_{diss}^{1/2}(1 + K_{G+}[G])^{1/2}/C_L^{1/2}$$

$$k_p = \{k_\pm + k_G' K_g'[G]\} + k_- K_{diss}^{1/2}(1 + K'_{G+}[G]^2)^{1/2}/C_L^{1/2} .$$

The first equation applies to the system involving tetraglyme, whereas the second applies to the solutions resulting from the addition of triglyme. The symbols k_G or k_G' denote the propagation constants of glymated ion-pairs, the others retaining their conventional meaning. The verification of these relations confirms the 1 : 1 stoichiometry of the associ-

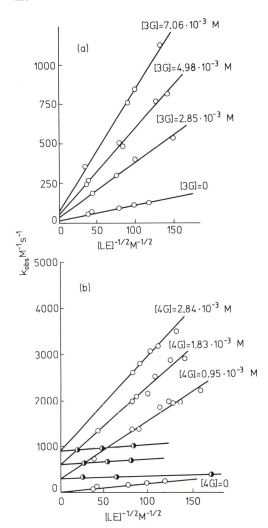

Fig. 53 a Plots of the observed bimolecular propagation constant, k_p, of sodium polystyrene in tetrahydropyrane at 25 °C vs. $1/C_L^{1/2}$ for various concentrations of added triglyme. **b** Plots of the observed bimolecular propagation constants, k_p, of sodium polystyrene in tetrahydropyrane at 25 °C vs. $1/C_L^{1/2}$ for various concentrations of added tetraglyme. The half-shaded points – the results obtained in the presence of an excess of tetraphenyl boride

ation of sodium ions with tetraglyme and the 1 : 2 stoichiometry of their association with triglyme*. The intercepts, I, of the lines shown in Figs. 53 should be linear with the glyme concentrations. This was shown to be the case as illustrated by Fig. 54, and the slopes of the resulting plots give $k_G K_G$. Thus, $k_G K_G$ was determined as 8.10^3 $M^{-2}s^{-1}$ and 3.10^5 $M^{-2}s^{-1}$ at 25 °C for the triglyme and the tetraglyme systems, respectively.

The square of the slopes of the lines k_p vs. $1/C_L^{1/2}$, denoted by S, were found to be linear with [G] for the tetraglyme systems, but the linearity required plotting k_p vs. $[G]^2$ for the triglyme system. This is shown in Figs. 55 from which the values of K_{G+} and K'_{G+} were determined.

---

* The 1 : 2 stoichiometry is established in the *investigated* range of triglyme. It results from a high degree of association of mono-glymated Na^+ with another triglyme molecule, making the concentration of the mono-glymated ions insignificant under those conditions

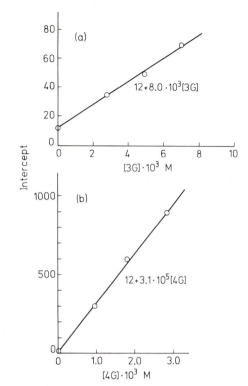

Fig. 54 a Intercepts of the lines shown in Fig. 53 a as a function of triglyme concentration. **b** Intercepts of the lines shown in Fig. 53 b as a function of tetraglyme concentration

Extension of this work to higher concentrations of tetraglyme[331)] led to determination of k_G and K_G. The pertinent experiments were performed in the presence of NaBPh$_4$ to eliminate the contribution of the free polystyryl anions to the propagation. The intercepts I's are given by the relation,

$$I = (1 - f)I_0 + fk_G ,$$

where f denotes the mole fraction of the "glymated" pairs and I_0 the intercept determined in the absence of glyme. Since $f/(1 - f) = K_G[G]$, one deduces the relation $1/(I - I_0) = 1/k_G + (1/k_G K_G)[G]^{-1}$. A plot of $1/(I - I_0)$ vs. $[G]^{-1}$ is shown in Fig. 56. It is linear, and from its slope and intercepts the values of k_G and K_G are derived. These are collected in Table 12. Alternatively, k_G is obtained by using excess of glyme, as shown in Fig. 57. The respective ΔH_G and ΔS_G are $- 6.5$ kcal/mol and 13 eu, while the activation energy $E_G = 1.2$ kcal/mol. Similar values of ΔH and ΔS were found the for glymation of other kinds of ion-pairs, viz.

Biphenylide$^-$, Na$^+$ + tetraglyme in THF,

$\Delta H = - 6.0$ kcal/mol, the association constant 75 M^{-1} at 21 °C[357)], and

Fluorenyl$^-$, Na$^+$ + tetraglyme in THF,

$\Delta H = - 7.0$ kcal/mol, the association constant 125 M^{-1} at 25 °C[358)].

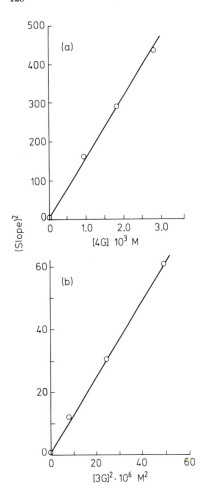

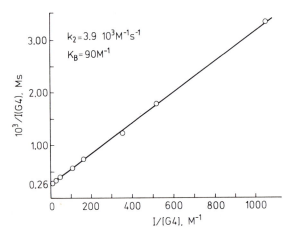

Fig. 55. a Square of slopes of the lines shown in Fig. 53 b as a function of tetraglyme concentration. **b** Square of slopes of the lines shown in Fig. 53 a as a function of *square* of concentration of triglyme

Fig. 56. Reciprocal of $\Delta I = k_{\pm} - k_{\pm,0}$ vs. reciprocal of tetraglyme concentration. $k_{\pm}$ and $k_{\pm,0}$ denote the propagation constants of sodium polystyryl ion-pairs in THP determined in the presence and absence of tetraglyme. T = 25 °C

Table 12. k_G and K_G determined for the system sodium polystyryl and tetraglyme in THF

T °C	25	0	− 20	− 45
k_g $M^{-1}s^{-1}$	3,900	3,500	2,600	2,100
K_G M^{-1}	90	210	770	2,000

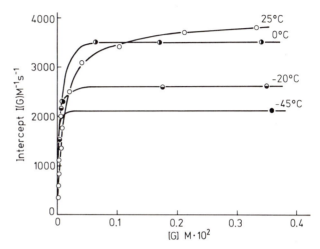

Fig. 57. Plot of $\Delta I = k_{\pm} - k_{\pm,0}$ vs. glyme concentration at various temperatures. The assymptotic values correspond to propagation constants of "glymated" sodium polystyryl ion-pairs in THP

The very low value of E_G seems to indicate the presence of two isomers, the less reactive form being more abundant at higher temperatures. Indirect evidence for this suggestion is given in Ref. 344.

Discussion of the effects caused by kryptates, a most powerful solvating agent, is given on p. 156.

IV.11. Propagation of Polydiene Salts in Ethereal Solvents

Studies of propagation of polydiene salts in ethereal solvents are hampered by their instability and by a variety of isomerization processes altering the structure of their active centers. These side reactions vitiated the early investigations of those systems and made the reports published prior to 1965 of little value. Stability of ethereal solutions of the diene salts is greatly improved when they are kept at low temperatures, especially in the presence of salts suppressing the dissociation of living polymers[398]*.

---

* In most systems, including living polystyrene, the free ions react faster with ethereal solvents than their ion-pairs

Participation of free ions and ion-pairs in propagation of lithium isoprene in THF was demonstrated by Bywater and Worsfold[393]. Since the reaction was investigated at 30 °C, when the annihilation processes are troublesome, the propagation constant, k_p, had to be determined from the initial rates. Nevertheless, the results led to a good linear relation of k_p vs. $1/C_L^{1/2}$, and addition of LiBPH$_4$ resulted in the expected suppression of the free ions' contribution to propagation. Thus, the propagation constant, $k_\pm$, of the lithium ion-pairs was determined as 0.2 M^{-1}s^{-1} and the slope of the linear relation, $k_-K_{diss}^{1/2}$, as $6 \cdot 10^{-2}$ M$^{-1/2}$s^{-1}, both at 30 °C. The kinetic work was supplemented by a conductance study yielding $K_{diss} = 5 \cdot 10^{-9}$ M. This value might be too high because some conducting products could be formed by the side reactions. If accepted, it leads to $k_- = 3 \cdot 10^4$ M^{-1}s^{-1}. The remarkably low K_{diss}, about 300 times smaller than that of lithium polysty-rene, implies a very tight structure of lithium polyisoprenyl in that solvent.

The existence of at least two distinct forms of the active ends of the polydienes was fully realized in the following years. These may acquire, e.g. a cis or a trans-configuration and their identification became feasible through spectroscopic and NMR studies[404, 405]. In view of these developments, the study of propagation of lithium and sodium polyiso-prenyl in THF was repeated and extended to include the spectroscopic data[400]. These revealed that some irreversible, as well as reversible, changes occur in that system. The products of the irreversible reactions were not identified, although tentative suggestions about their structure have been made. The reversible changes were attributed to the cis-trans-isomerization. The cis-form of the lithium polyisoprenyl is preferred in THF at low temperatures; however, the trans-form is preferentially formed on the monomer addition and it reverts slowly into the cis-isomer after cessation of polymerization. This is man-ifested by an abrupt change of the electronic spectrum of an equilibrated solution of the salt on addition of monomer, followed by a slow reappearance of the original spectrum after completion of polymerization. For the reader's sake, the constructed spectra of the cis and trans-isomers are shown in Fig. 58.

The high reactivity of the free polyisoprenyl ion was confirmed; the polymerization proceeds much faster in the absence than in the presence of LiBPh$_4$. The propagation

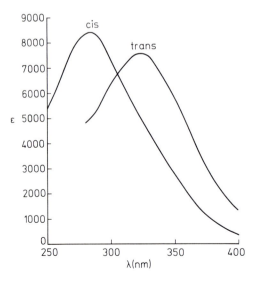

Fig. 58. The spectra of the *cis*- and *trans*-forms of sodium polybutadienyl in THF

constant, $k_{\pm}$, of the lithium ion-pairs (preferentially trans?) was determined over a wide range of temperatures. The Arrhenius plot was linear, the activation energy being 10 kcal/mol, the A = $4 \cdot 10^6\,M^{-1}s^{-1}$. The considerable instability of the sodium salt made the interpretation of the results obtained in that system difficult and unreliable.

The cis-form of lithium polyisoprenyl is also preferred in diethyl ether solution[401]. The spectroscopic and NMR studies were supplemented by a chemical fixation of the structure of a model compound; addition of trimethyl silyl chloride yields adducts (presumably without isomerizing the reacting salt) which were eventually separated by a chromatographic technique into cis- and trans-isomers. The results confirmed the previous NMR analysis.

The electronic spectrum of lithium polyisoprenyl in diethyl ether changes at higher concentration of the salt, $> 10^{-3}$ M. Apparently the ion-pairs associate in this medium into higher aggregates and the concomital decrease of their reactivity implies a lower reactivity of the aggregates, a conclusion made previously by Sinn[406].

Studies of the sodium salts were prevented by technical difficulties. However, reproducible data were reported for the potassium salt. Its propagation is more than 20 times as fast as that of the lithium salt, the increase is caused by lower activation energy, viz. about 12 kcal/mol for the Li$^+$ salt compared with 9 kcal/mol and A = $5.10^6\,M^{-1}s^{-1}$ for the potassium salt. This gradation seems to be typical for the diene salts; it indicates their tight structure that demands a higher rate for larger counter-ions.

Propagation of polybutadiene salts in THF was investigated by Garton and Bywater[394 a] and by Sigwalt et al.[399]. The side reactions again hampered these studies, although to a lesser extent than in the isoprene system. The former group followed the propagation by a spectrophotometric technique, studying the Li$^+$, Na$^+$ and K$^+$ salts, while the work of the latter group was limited to the potassium salt only and utilized a calorimetric technique. A good linear plot of k_p vs. 1/[Na$^+$] was reported for polybutadienyl sodium at $-70\,°C$, the concentration of sodium ions being varied by the addition of NaBPh$_4$. No attempt was made to determine accurately the K$_{diss}$ and k_-; however, from the results of a single experiment performed in the absence of the boride, K$_{diss}$ was estimated as $\sim 10^{-9}$ M and k_- as $\sim 10^4\,M^{-1}s^{-1}$ at $-70\,°C$. Hence, at 10^{-3} M concentration of living polymers, the free ions represent 0.1% of the growing species but they contribute more to the reaction than the ion-pairs. The reactivity of the sodium ion-pairs is very low at that temperature, $\sim 0.3\,M^{-1}s^{-1}$, while the contribution of the free ions amounts to $\sim 10\,M^{-1}s^{-1}$. The Arrhenius plot of $k_{\pm}$ shown in Fig. 59 is strongly curved, the apparent activation energy decreasing with temperature. This observation could indicate a gradual conversion at lower temperatures of the less reactive tight pairs into the more reactive loose pairs, like e.g. in the styrene system. However, in view of their spectroscopic data[394 b], the authors perfer to interpret their results in terms of the cis-trans-isomerization; decreasing temperature increasingly favors the more reactive trans-form.

The last point needs elaboration. The spectroscopic data[394 b] pertaining to *equilibrated* solutions of the lithium or sodium salts indicate an increase in the cis content at lower temperatures. However, the propagation yields preferentially the trans-form, apparently even to a larger degree at lower temperatures. This situation might be described in terms of four propagation steps: addition of the monomer to the cis-end of a growing polymer yields either a new cis-terminated polymer, the rate constant k_{cc}, or a trans-one, with the rate constant $k_{c,t}$. Analogous two steps that convert a trans-termi-

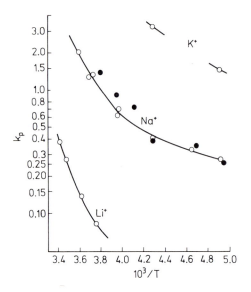

Fig. 59. Arrhenius plots of propagation constants, $k_\pm$, of polybutadiene ion-pairs in THF for Li$^+$, Na$^+$, and K$^+$ salts. The *full circles* represent the results of Funt derived from electrochemical studies

nated polymer are governed by the constants k_{tc} and k_{tt}. Apparently $k_{ct} > k_{cc}$ and the ratio k_{ct}/k_{cc} increases with decreasing temperature, while $k_{tt} > k_{tc}$.

The lithium pairs are substantially less reactive and less dissociated than the sodium salt, the pertinent results being displayed in Fig. 59. The Arrhenius line seems to be straight and corresponds to activation energy of ~ 10 kcal/mol with A $\sim 10^7$.

Two sets of data were reported for the propagation of potassium polybutadienyl in THF. Those of Sigwalt et al.[398] were derived from the dependence of the observed propagation constant k_p on the concentration of living polymers (linearity with $1/C_L^{1/2}$) and on the concentration of K$^+$ ions, the latter being varied by the addition of KBPh$_4$ (linearity with $1/[K^+]$). They led to the following values of K_{diss}, k_- and $k_\pm$:

T °C	$10^9 K_{diss}$ in M	$10^4 k_-$ in M^{-1}s^{-1}	$k_\pm$ in M^{-1}s^{-1}
0	7.8	4.8	<1 (?)
−6	9.4	3.6	<1 (?)
−13	11.0	2.6	<1 (?)

$E_- = 6.5$ kcal/mol; $A_- = $ M^{-1}s^{-1}.

The results of Garton and Bywater[394 a] were derived from only two experiments. The rates were determined in the presence of KBPh$_4$ and were attributed to propagation by the ion-pairs, viz. $k_\pm$ of ~ 3 M^{-1}s^{-1} at -40 °C and ~ 1.5 M^{-1}s^{-1} at -70 °C, both values higher than Sigwalt's findings.

An interesting method leading to determination of propagation constants of living polymers was developed by Funt[395]. Polymerization of that monomer was initiated in an electrolytic cell by a pulse of reducing current (i.e. through an electron-transfer technique). The resulting living polymers grow for a predetermined period and then are terminated by a pulse of oxidizing current. The concentration of living polymers is calculated from the Coulombmetric data, and the amount of polymerized monomer then allows one to calculate the propagation constant. This approach was applied to polymerization of butadiene in THF in the presence of a large excess of $NaBPh_4$ serving as a supporting electrolyte. The respective values of $k_\pm$ for the sodium salts are displayed in Fig. 59 (the filled circles) and show a reasonable agreement with those obtained by Garton and Bywater.

IV.12. Anionic Polymerization of Polar Vinyl Monomers

Anionic polymerization of methyl methacrylate is the most intensively studied polymerization of polar monomers. The early evidence for its feasibility was provided in 1956 by Rembaum and Szwarc[361], who initiated anionic polymerization of that monomer in THF at $-78\,°C$ with sodium naphthalenide or sodium polystyryl. The ensuing polymerization was found to be quantitative, completed within seconds, but the resulting polymer appeared to be terminated within half an hour. The termination was ascribed to the reaction $\sim\!\!\sim CH_2\overline{C}(CH_3)CO\cdot OCH_3 \rightarrow \sim\!\!\sim CH_2C(CH_3)_2\cdot COO^-$. However, in view of the now available evidence[360, 382], it seems likely that some impurities still remaining in the system destroyed the living polymers.

Shortly afterward, the discovery of stereospecific, crystalline polymethyl methacrylate, reported in 1958 by Fox et al.[362], greatly stimulated interest in anionic polymerization of methyl methacrylate. A continuation of that work by Glusker and co-workers[363] demonstrated the living character of anionic polymerization of this monomer when initiated by lithium fluorenyl in toluene within a temperature range of -50 to $-78\,°C$, (see also[382]). Initiation appeared to result from the addition of fluorenyl carbanion to the $C=C$ double bond of the monomer, a reaction virtually completed in 5 s, while the lack of termination was demonstrated by the formation of radioactive polymers on quenching the reaction with radioactive CO_2 or tritiated acetic acid. However, the apparent simplicity of this polymerization became questionable because Glusker et al. also reported a very broad molecular weight distribution of the product, in contrast to the expected Poisson distribution.

An increasing number of puzzling and often contradictory features of this polymerization was revealed by the subsequent studies. These were reviewed in 1965 by Bywater[364], who demonstrated that none of the mechanisms proposed by the various investigators is capable of accounting for all the observations. Although many of the reported complexities are still not elucidated, some of them are at least partially understood.

Much of the difficulties arising in the early investigations resulted from the unfortunate choices of experimental conditions. Use of organolithium compounds as the initiators is troublesome, especially in mechanistic studies, because they react with methyl methacrylate in more than one fashion*. Denothing the organo-lithium reagent by RLi, we may visualize three modes of interactions:

---

* Some difficulties could be avoided by using initiators such as e.g. 1,1-diphenyl hexyl lithium

$$RLi + CH_2:C(CH_3)CO \cdot OCH_3 \longrightarrow \begin{array}{l} RCH_2 \cdot \bar{C}Li^+(CH_3)CO \cdot OCH_3 \\[4pt] CH_2:C(CH_3) \cdot CO \cdot R + CH_3OLi \\[4pt] CH_2:C(\bar{C}H_2Li^+)CO \cdot OCH_3 + RH \end{array}$$

The second mode is especially damaging. It causes not only a loss of the initiator but yields lithium alkoxide which, in turn, interferes with the propagation. In this respect butyl lithium is the most undesired initiator as it produces large amounts of the alkoxide in the first few seconds even at $-30\,°C$. Similar side reactions take place in the propagation of lithium polymethyl methacrylate, causing, e.g. ring formation or a "false" addition and branching[371, 372].

$$\underset{\underset{CO \cdot OCH_3}{|}}{\overset{\overset{CH_3}{|}}{\text{ww} CH_2 \cdot \overset{|}{C}^-}}, Li^+ \;+\; CH_2:C(CH_3) \cdot CO \cdot OCH_3 \longrightarrow$$

$$\underset{\underset{CO \cdot OCH_3}{|}}{\overset{\overset{CH_3}{|}}{\text{ww} CH_2 \cdot \overset{|}{C}}}-CO \cdot C(CH_3):CH_2 \;+\; Li^+,\, {}^-OCH_3, \text{ false addition,}$$

$$\underset{\underset{CO \cdot OCH_3}{|}}{\text{ww} CH_2 \cdot C(CH_3)\text{ww}} \;+\; Li^+,\, \bar{C}(CH_3)(CO \cdot OCH_3)\text{ww} \longrightarrow$$

$$\underset{\underset{CO \cdot C(CH_3)(CO \cdot OCH_3)\text{ww}}{|}}{\text{ww} CH_2 \cdot C(CH_3)\text{ww}} \;+\; Li^+,\, {}^-OCH_3, \text{ branching}$$

Use of toluene or toluene-ether mixtures as solvents further contributed to the complexity of the process since appreciable aggregation of growing polymers seems to occur in these media. Use of fluorenyl salt yields polymers possessing a terminal fluorenyl moiety. Its acidic proton terminates a growing polymer simultaneously with formation of the respective carbanions[373]. The latter initiates further polymerization resulting in formation of a polymer endowed with two growing end-groups, thus favoring intra-molecular aggregation[375]. In view of all these observations, it became obvious that progress in this field required development of "clean" systems in which the side reactions and the undesirable complexities are eliminated or at least minimized.

Formation of polymers with relatively narrow molecular weight distribution provides a fair evidence for the "cleaness" of the system. This was achieved in polymerization of meticulously purified methyl methacrylate initiated by biphenylide salts of alkali metals and carried out in polar solvents, THF or DME, at a low temperature, $-78\,°C$[360]. Under those conditions, truly living polymethyl methacrylates were prepared and their stability made it possible to investigate their state of aggregation and their conductance[368]. By comparing the viscosity of living polymethyl methacrylate solution and that of the terminated, say protonated one, Figueruelo showed an absence of aggregation in 10^{-4} M solutions. However, this conclusion is questioned, even by the author, since the investigated polymers possessed two growing end-groups and the aggregation could be intramolecular. The conductance studies[368] revealed the extremely low degree of dissociation of alkali salts of living polymethyl methacrylate, K_{diss} being estimated at $\sim 10^{-10}$ M for the Li^+ and K^+ salts in THF at $-78\,^\circ$C.

The work of Figueruelo was confirmed and extended by Schulz' group[365] and by Mita et al.[370]. Conductance studies of cesium salt of polymethyl methacrylate, initiated in THF by cumyl cesium or cesium salt of α-methyl styrene oligomers, led to $K_{diss} = 2 \cdot 10^{-9}$ M at $-78\,^\circ$C, confirming its low degree of ionization[366]. Similar results were reported for the sodium salt, viz. $\sim 3 \cdot 10^{-9}$ M both at $-40\,^\circ$C and at $-78\,^\circ$C[369]. The cesium salt of living polymethyl methacrylate probably was mono-functional and therefore the reported K_{diss}'s were reliable. On the other hand, bifunctional polymers were utilized in preparation of the sodium salt; and as has been shown later[375], these undergo *intra*-molecular association, especially at lower degree of polymerization. It is desirable, therefore, to recheck these results by determining conductance of solutions of monofunctional sodium polymethyl methacrylate.

Kinetic results reported by Löhr and Schulz[365, 369] appeared to be reliable and self-consistent. The resulting polymers showed the expected narrow molecular weight distribution, the rate of monomer consumption seemed to obey first order law, and the number average degree of polymerization increased proportionally with the degree of conversion. All these observations confirmed the living character of the polymerization at low temperatures, implying the absence of termination and of other side reactions. Since the bimolecular propagation constant, k_p, derived from that work appeared to be linear with $1/C_L^{1/2}$, the propagation constants of the free polymethyl methacrylate anions, k_-, and of their ion-pairs, $k_\pm$, were calculated by the conventional procedure. The k_- values derived from the experiments performed with the cesium and sodium salts at $-75\,^\circ$C agreed with each other and were reported as $4.8 \cdot 10^4\,M^{-1}s^{-1}$ and $4.5 \cdot 10^4\,M^{-1}s^{-1}$, respectively. The $k_\pm$ constants for the cesium salt were determined at several temperatures, e.g. $700\,M^{-1}s^{-1}$ at $-30\,^\circ$C and $80\,M^{-1}s^{-1}$ at $-75\,^\circ$C, leading to activation energy of 4.5 kcal/mol. Some lower values were reported for the sodium salt.

Unfortunately, subsequent reports made these results questionable. Two deficiencies were revealed in the work of Löhr and Schulz[373]. At the usual concentration range of monomer and initiator, the polymerization was too fast for the application of the batch technique but too slow for a reliable utilization of the flow technique. Indeed, some broadening of molecular weight distribution was observed and it was attributed to laminarity of the flow. Moreover, experiments were performed with salts of bifunctional polymethyl methacrylate that undergo *intra*-molecular association, especially at the early stages of polymerization[375]. Such an association, as will be shown later, affects the kinetics of propagation and distorts its apparent first order character.

Indeed, improvement of the technique led to about twice as large values of $k_\pm$ of the cesium ion-pairs as reported by Löhr and Schulz[365, 369], namely $\sim 900\,M^{-1}s^{-1}$ at $-50\,^\circ$C, and $35\,M^{-1}s^{-1}$ at $-100\,^\circ$C, yielding the respective $E_\pm = 4.9$ kcal/mol and log $A_\pm = 7.7$[379]. The previously reported $k_\pm$ values of sodium polymethyl methycrylate were shown to be of little significance because they refer to bifunctional polymethyl methacrylate, or rather to a mixture of mono-functional ($\sim 10\%$?) and bifunctional forms.

The problem of intra-molecular association of sodium salt of bifunctional polymethyl methacrylate came clearly into focus in the report of Warzelhan and Schulz[375]. By using the improved technique described elsewhere[266], they showed that the consumption of the monomer deviated from the expected first order reaction – the propagation became faster as conversion progressed. This is shown by Fig. 60. Nevertheless, the number average degree of polymerization increased proportionally with the degree of conversion, demonstrating that the acceleration was not caused by a slow initiation. These nonconventional results were explained by postulating intra-molecular association of the active end-groups; its extent decreases as the chain linking the interacting end-groups becomes longer, an effect observed earlier by Bhattacharyya et al.[264 b, 295] and predicted for other systems by Szwarc[376]. Furthermore, it was postulated that the associated end-groups propagate slower (per active end-group) than the unassociated.

Additional complexity of the studied polymerization arose from the presence of some mono-functional polymers formed through partial destruction of bifunctional initiators*. This phenomenon was observed and accounted for in the early work on anionic polymerization, e.g.[377]; it leads to bimodal molecular weight distribution of the resulting polymer as noted by other investigators, e.g.[360]. The bimodal molecular weight distribution was effectively utilized by Warzelhan and Schulz to confirm the existence of two kinds of growing species and to demonstrate the sluggishness of propagation by the *intra*-molecularly associated polymers.

The partial destruction by impurities of bifunctional initiator or dianionic oligomers is equally probable in other anionic polymerizations, e.g. in the polymerization of styrene. Even in the absence of intra-molecular association, such a destruction should be revealed by a deviation of the molecular weight distribution of the resulting polymer from the Poisson distribution. Significantly, in spite of the use of bifunctional initiators, disturb-

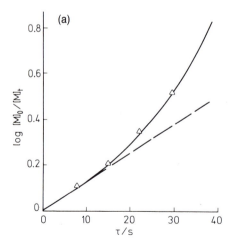

Fig. 60. Plot of log $[M]_0/[M]_t$ vs. time for polymerization of sodium salt of living polymethyl methacrylate prepared in THF through initiation by bifunctional initiator

---

* Or by partial destruction of bifunctional polymers in the early stages of reaction by impurities still left in the system

ances of that kind, which could nullify some of the conclusions of Schulz' group, were not observed in their studies of anionic polymerization of styrene.

The formation of mono-functional living polymethyl methacrylate in a process initiated by bi-functional initiators and performed under the most stringent conditions of purity could perhaps be specific for that monomer and not caused by impurities. The results of experiments performed with long, bifunctional oligomers of α-methyl styrene[374] are most significant. Although the auto-acceleration was eliminated, the molecular weight distribution of the polymers still showed a bimodal character. It may be that ~ 10% of the monomer reacts with carbanions of α-methyl styrene or the di-anions of the dimeric methyl methacrylate in a destructive way; however, such a reaction does not occur in the subsequent propagation. This is a highly speculative suggestion and the problem deserves further investigation.

A recent paper by Chaplin et al.[378] should be noted. These workers explicitly contradict Schulz' findings and claim a lack of bimodal distribution in their polymers resulting from initiation by bi-functional initiators. Nevertheless, this writer does not doubt that mono-functional polymers were formed in their system as evident by the reported dependence of $\overline{DP}_w/\overline{DP}_n$ on the degree of conversion, by the inequality $\overline{DP}_n$(observed) < $\overline{DP}_n$(calculated), and by the values of the reported rate constants.

Studies of the different behavior of mono- and bi-functional living polymethyl methycrylate were reported by Höcker and co-workers[373, 374]. Sodium salt of mono-functional living polymethyl methacrylate was prepared in THF at low temperature by utilizing the adduct of benzyl sodium to α-methyl styrene as an initiator,

$$C_6H_5 \cdot CH_2\{CH_2 \cdot C(CH_3)(Ph)\}_n CH_2 \cdot \overline{C}(CH_3)(Ph)Na^+ .$$

The experiments were performed in the improved reactor[266] that eliminates the shortcomings of Löhr and Schulz' work, and the contribution of free anions was suppressed by the addition of NaBPh$_4$. The results manifested a high degree of reliability of that system – free of any perturbance. The propagation constants of sodium ion-pairs, $k_\pm$, were determined at temperatures ranging from $-50\,°C$ to $-100\,°C$, namely $k_\pm = 440\ M^{-1}s^{-1}$ at $-50\,°C$ and $k_\pm = 30\ M^{-1}s^{-1}$ at $-100\,°C$. The Arrhenius plot was perfectly linear over a wide temperature range, the activation energy $E_\pm = 4.7$ kcal/mol and $\log A_\pm = 7.0$. The linearity of the Arrhenius plot, contrasting the curvature observed in the polystyrene system, suggests that only one kind of ion-pairs participates in the polymerization of mono-functional sodium-polymethyl methacrylate in THF. These pairs cannot be described as ordinary contact pairs since their reactivity is remarkably similar to that of the cesium pairs[379]. Moreover, the $k_\pm$'s are large while the respective K_{diss}' are exceedingly small. Apparently these ion-pairs are *intra*-molecularly solvated by the polar group of the penultimate or antepenultimate segments of the polymer, a structure suggested by Schuerch et al.[380] on the basis of the NMR spectra.

In this respect, the ion-pairs of living polymethyl methacrylate resemble those of sodium poly-2-vinyl pyridine[352, 353]. The *intra*-molecular solvation prevents the formation of freely moving cations but exposes the carbanions and enhances their reactivity.

Only one kind of *intra*-molecular complexation is possible for the salts of poly-2-vinyl pyridine or mono-functional polymethyl methacrylate, namely that depicted here or on p. 116. Replacement of Na$^+$ by Cs$^+$ distorts it; Cs$^+$ being too large does not fit into the cavity and therefore the Cs$^+$ ion-pairs may dissociate easier than the Na$^+$ pairs[354]. Nevertheless, this different placement only slightly affects the reactivity of the poly-methyl methacrylate pairs because cations are located relatively far away from the reactive $\bar{C}$ center, being strongly associated with oxygen of the carbonyl group.

Intra-molecular solvation of sodium pairs of poly-2-vinyl pyridine accounts for the virtual independence of their reactivity on the solvent's nature. The large variation of $k_\pm$'s observed in the styrene system on changing solvents results from concomitant variation of the ratio [tight pairs]/[loose pairs], while such an isomerism does not occur in the intra-molecular complexes. In fact, larger values of $k_\pm$ were reported for the polymerization of 2-vinyl pyridine proceeding in dioxane or tetrahydropyrane than for the reaction in THF[353]. This observation was rationalized by stressing the necessity of replacing a neighboring solvent molecule by a monomer to allow for propagation, and the stronger solvation, the more difficult the replacement[353]. One might expect a similar behavior of mono-functional polymethyl methacrylate. However, in a powerfully solvating medium, e.g. in DME, ion-pairs of polymethyl methacrylate may acquire a new form[383] viz.

These pairs should be treated as very tight enolate ion-pairs, a suggestion supported by the results of the ^{13}C-NMR studies of model compounds, e.g. of metalated methyl isobuterate[386]. Hence, their dissociation is negligible but their reactivity is higher than that of the previously discussed carbanionic pairs, due to the increased exposure of the carbon center[383, 384]. Indeed, the respective $k_\pm$'s are substantially larger than those found in THF, e.g., $\sim 2,100$ M^{-1}s^{-1} for the Na$^+$ and $\sim 1,700$ M^{-1}s^{-1} for the Cs$^+$ pairs in DME at $-50\,°$C, compared with 440 M^{-1}s^{-1} for either salt in THF at the same temperature. The remarkable indifference of the reactivity to the nature of the counter-ion (Na$^+$ or Cs$^+$) reflects again the remoteness of the cation from the carbon reaction center.

The behavior of bifunctional sodium polymethyl methacrylate greatly differs from that of the mono-functional polymer. The polymerization is auto-accelerated, as reported by Warzelhan and Schulz[375] and confirmed by Höcker et al.[379]. The evidence presented in both papers conclusively shows that this phenomenon arises from an increase in propagation constant on increasing the length of the chains, and not from an increase of the concentration of active centers. Indeed, the latter concentration remains constant in each run. As pointed out earlier the equilibrium between associated and separated end-groups is most plausible, viz.

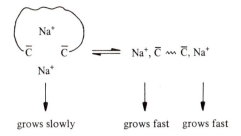

grows slowly   grows fast   grows fast

It accounts for the observations since the equilibrium shifts to the right as the chain becomes longer. Further evidence in its favor is provided by the results derived from the polymerizations initiated by large, bifunctional oligomers of α-methyl styrene (DP 70 and 270)[374]. In such a polymerization the two $-\overline{C}(CH_3)CO \cdot OCH_3$ groups are far apart from the very start of the reaction, hence their association is prevented. Indeed, no auto-acceleration was observed in such runs.

By combining the kinetic and fractionation data, the authors calculated the propagation constant of associated ends to be 27 $M^{-1}s^{-1}$ (per active end). Using a different approach this writer concluded that the respective constant is smaller than 15 this $M^{-1}s^{-1}$. Both evaluations reveal a substantial decrease of reactivity arising from the association.

A question arises whether this low rate of propagation characterizes the rate of addition to one of the groups still associated with the other, or if it results from "normal" fast addition to unassociated pairs which are at low concentration in equilibrium with the associated ones. The latter situation was observed in the propagation of living polystyrene taking place in the presence of an excess of anthracene[385]. The adduct, denoted by $\sim \overline{S}, A$, does not propagate, but it remains in equilibrium with the uncomplexe polystyrene and anthracene; namely

$$\sim \overline{S}, A \;\rightleftharpoons\; A + \sim \overline{S}$$

does not   propagates .
propagate

The equilibrium lies far to the left and the observed rate of polymerization was found to be inversely proportional to the concentration of the uncomplexed anthracene – an evidence confirming the proposed mechanism.

Unfortunately, this kind of discrimination is unfeasible in the polymethyl methacrylate system. However, the tacticity of the polymer provides a useful way of differentiating between these two alternatives. The polymethyl methacrylate formed in THF by monofunctional initiators is, to a high degree, syndiotactic. The tacticity should have remained unaltered had the association-dissociation mechanism been operative, whereas the polymer formed in the early stages of a reaction initiated by bifunctional initiators shows a larger proportion of isotactic diads. This demonstrates that the slow addition involves a different center, i.e. it takes place on the associated end-groups. A further discussion of tacticity is postponed here and will be continued later.

Polymerization of methyl methacrylate initiated by Grignard compound is most complex[378, 387] and it is premature to discuss this topic in view of a paucity of information. Similarly, complex phenomena, requiring more investigation, obscure the mechanism of its polymerization in nonpolar solvents or in mixtures of polar and non-polar media.

Undoubtedly, the aggregation becomes tighter and more extensive as the proportion of nonpolar component increases. Such an aggregation favors the isotactic placement but unfortunately enhances the side reactions, resulting in the formation of alkoxides and ring compounds.

IV.13. Propagation of Styrene and the Dienes Polymerization in Hydrocarbon Solvents with Li^+ Counter-Ions

The early studies of those reactions, e.g.[157, 158] were confused by lack of differentiation between propagation and initiation and by the meager knowledge of the association phenomena that play an important role in those processes. The mechanism of propagation of lithium polystyrene in benzene was elucidated by the pioneering work of Worsfold and Bywater[156]. As stated previously, the use of spectrophotometric techniques allowed them to independently follow the course of propagation and of initiation and to show the proportionality of the propagation rate with the square-root of living polymers concentration. This observation was accounted for by the following mechanism:

$$(\text{\textasciitilde} S^-, Li^+)_2 \rightleftharpoons 2(\text{\textasciitilde} S^-, Li^+) , \qquad K_{eq} ,$$

$$\underset{\text{n-mer}}{\text{\textasciitilde} S^-, Li^+} + S \rightarrow \underset{\text{n + 1-mer}}{\text{\textasciitilde} S^-, Li^+} , \qquad k_p ,$$

where $\text{\textasciitilde} S^-, Li^+$ denotes lithium polystyrene and S, a styrene molecule. At ambient temperatures the equilibrium of the dimerization lies far to the left; at least 90% of growing polymers are dimeric, even at 10^{-5} M concentration of lithium polystyrene, implying $K_{eq} \leq 10^{-7}$ M. The alternative interpretation of the square-root relation, namely,

$$\text{\textasciitilde} S^-Li^+ \rightleftharpoons \text{\textasciitilde} S^- + Li^+ , \qquad K_{S^-Li^+} ,$$

is unlikely. Bare Li^+ cations could not be present in benzene at any detectable concentration, and indeed, conductance studies showed the absence of free ions in those solutions, i.e. $K_{S^-,Li^+} < 10^{-12}$ M.

The reality of dimerization was soon confirmed by comparing the viscosities of living lithium polystyrene solutions with those of the terminated polymers[388] *, a finding verified later by alternative techniques[173, 392]. The kinetic data led to $k_p K_{eq}^{1/2} = 1.6 \cdot 10^{-6}$ $M^{-1/2}s^{-1}$ at 30 °C and therefore $k_p > 50$ $M^{-1}s^{-1}$. This is a high value for the propagation constant of tight ion-pairs, suggesting the involvement of a push-pull mechanism in the monomer addition. Apparently, the oncoming monomer becomes associated with Li^+ cation and is strongly polarized prior to its addition. Unfortunately, this suggestion cannot be tested by compatitive experiments because the hypothetic complex represents presumably only a minute fraction of living polymers, i.e. the equilibrium

$$\text{\textasciitilde} S^-, Li^+ + S \rightleftharpoons \text{\textasciitilde} S^-, Li^+, S$$

lies far to the left.

---

* The shortcomings of this technique have been demonstrated recently[419]

Let us digress and ask whether $\sim$ S$^-$, Li$^+$ in benzene should be treated as a tight ion-pair or as a covalently linked Li to a benzylic carbon. An attempt to clarify the meaning of such a question was discussed elsewhere[323 c, 449]. However, what needs stressing is the similarity of the electronic spectrum of lithium polystyrene in benzene to that of its THF solution, or for that matter, to the spectrum of sodium polystyrene in THF[389]. The ionic structure of those salts in THF is beyond any doubts*. This observation should be contrasted with the findings of greatly different spectra of the ionic and covalent forms reported for other systems, e.g. of the covalently bonded Ph$_3$C · Cl in benzene and the ionic Ph$_3$C$^+$, Cl$^-$ in nitromethane.

The kinetic results of Worsfold and Bywater were confirmed by the studies of lithium polystyrene propagation in toluene[390]; virtually identical rate constants were reported in both investigations. The propagation taking place in cyclohexane is slower by about a factor of 3[173]; nevertheless, its dependence on the square-root of living polymer concentration was confirmed, implying that the same mechanism of propagation is valid for the solution in aliphatic cyclohexane as for that in the aromatic benzene**.

Kinetics of propagation of lithium polyisoprenyl or butadienyl resemble that of lithium polystyrene. The logarithmic plots of the rate vs. living polymer concentration are again linear; however, their slopes are smaller, about ¼ for the isoprene system[174, 178, 390, 391] and perhaps even less for the butadiene system[173, 390]. Differences between low values of the slopes, e.g. the distinction of ¼ from ⅙, are within experimental uncertainties. Furthermore, deviations are sometimes observed when the concentration of living polymers is lesser than 10^{-4} M; presumably some effects caused by impurities become significant at high dilution. One group of investigators claimed a value ½ for the slope pertaining to the isoprene system, but they admitted in their later report that impurities vitiated their earlier results and verified the ¼ value.

These low values of slopes suggest a higher degree of aggregation for lithium polyisoprene or polybutadiene than for lithium polystyrene. Aggregation to tetramers was observed by the viscosity and light scatter methods for solutions of lithium polyisoprene[392]. Perhaps the most convincing evidence for a higher degree of association in hydrocarbon media of lithium polybutadiene or polyisoprene than of polystyrene was provided by the recent results of Hsieh[468]. By using an electronic device inserted into the studied liquid, Hsieh measured the viscosity of hydrocarbon solutions of lithium polystyrene. The device recorded a substantial increase of the viscosity when small amounts of butadiene or isoprene were added, amounts too small to measurably alter the molecular weight of the investigated polymers. However, such additions converted the terminal $^-$CH$_2$ · CH(Ph)Li groups into lithium butadienyl or isoprenyl end-groups. Subsequent judicious addition of small amounts of styrene reformed the original lithium polystyryl end-groups, and the device then recorded a drop of viscosity to its previous value. These results unequivocally demonstrate a higher degree of association when the end-groups of lithium polystyrene are converted into lithium butadienyl or isoprenyl.

Such results were reported earlier, e.g.[392], contrary to still earlier claims by other workers of lack of any change of viscosity in such experiments. However, closer inspection of some of their published results[473] casts serious doubts on the reliability of their

---

* Recent ab initio calculations by Streitwieser imply that even LiCH$_3$ is ionic
** Mechanism of initiation of polymerisation by alkyl lithiums in aromatic hydrocarbon is entirely different from that operating in aliphatic hydrocarbon, see p. 64

measurements[396, 474]. Other examples of questionable claims (e.g. of the dimeric nature of the aggregates) are discussed elsewhere[171, 397].

A claim of dimeric aggregation in conjunction with the established ¼ law of propagation seems to be self-contradictory. One should conceive a mechanism that would reconcile these apparently incompatible observations. None was proposed by the advocates of the dimeric aggregation. This writer suggested therefore a possible way out[397]. The mechanism assumes the dimeric nature of the aggregates and postulates that free polymer anions, P^-, are the only propagating species, i.e. denoting the monomer by M, the reaction

$$P^- + M \rightarrow P^- , \quad k_p$$
$$n\text{-mer} \qquad n + 1\text{-mer}$$

is the only one that contributes to polymerization. Of course, the simple idea of dissociation of the dimer, $(P^-, Li^+)_2$, into free ions, namely

$$(P^-, Li^+)_2 \rightleftharpoons 2P^- + 2Li^+ ,$$

is unrealistic. The presence of free lithium cations in hydrocarbon media is inconceivable, although the free polymeric P^- ions could be visualized. Hence, an alternate mode of ions formation is proposed:

$$(P^-, Li^+)_2 \rightleftharpoons (P^-, 2Li^+) + P^- , \quad K_2 .$$
$$\text{a positive}$$
$$\text{triple ion}$$

Two additional equilibria are considered:

$$(P^-, Li^+)_2 \rightleftharpoons 2(P^-, Li^+) , \qquad K_1$$
$$P^- + (P^-, Li^+) \rightleftharpoons (2P^-, Li^+) , \qquad K_3 .$$
$$\text{a negative}$$
$$\text{triple ion}$$

On solving the equations representing these equilibria, subject to the condition of electric neutrality of the system, one gets the relation

$$(\text{Rate of propagation})/[M] = k_p K_2^{1/2} [(P^-, Li^+)_2]^{1/2} / \{1 + K_1^{1/2} K_3 [(P^-, Li^+)_2]^{1/2}\}^{1/2} .$$

Since K_1 is a very small, while K_3 is expected to be large, one may conceive two extreme cases, namely, $K_1^{1/2} K_3 [(P^-, Li^+)_2]^{1/2} \gg 1$ and $K_1^{1/2} K_3 [(P^-, Li^+)_2]^{1/2} \ll 1$. The first extreme leads to

$$(\text{Rate of propagation})/[M] \sim [(P^-, Li^+)_2]^{1/4}$$

whereas the other demands

$$(\text{Rate of propagation})/[M] \sim [(P^-, Li^+)_2]^{1/2} .$$

The attractive feature of this mechanism is its ability to account for either ½, ¼, or any in between order of propagation of the presumably dimeric living polymers. Hence, it is applicable to the styrene as well as to the isoprene systems. Nevertheless, this writer has reservations about its validity because the stereospecific polymerization of lithium isoprene seems to demand the presence of Li^+ cation in the transition state of propagation.

The proposed mechanism could be tested. It predicts a change in the reaction order from ½ at low concentrations of living polymers to ¼ at their higher concentrations, i.e. the inequality

$$K_1^{1/2} K_3 \, [\text{living polymers}]^{1/2} \ll 1 \, ,$$

valid at low concentrations, might be reversed at sufficiently high concentrations resulting in

$$K_1^{1/2} K_3 \, [\text{living polymers}]^{1/2} \gg 1 \, .$$

Thus, the ½ order observed for the lithium polystyrene propagation might change into ¼ at some sufficiently high concentrations of living polystyrene, or the ¼ order observed for lithium polyisoprenyl should become ½ on their appropriate dilution. However, there are limitations and technical difficulties which could make such a test at least difficult if not unfeasible.

The aggregation of living lithium salts is the cause of some peculiar kinetic phenomena observed in a few systems. For example, the addition of 1,1-diphenyl ethylene, D, to lithium polystyrene, $\overline{S}, Li^+$, in benzene obeys the first order kinetics in living polymers, in spite of their dimeric nature. Moreover, at constant concentration of D, present in excess, the first order constant is inversely proportional to square-root of the initial concentration of living polystyrene[402]. Lithium polystyrene absorbs at $\lambda_{max} = 334$ nm while the adduct, $\sim CH_2CH(Ph) \cdot CH_2\overline{C}(Ph)_2, Li^+$ denoted by $\sim S\overline{D}, Li^+$, absorbs at $\lambda_{max} = 460$ nm. Hence, the progress of the reaction is readily followed spectrophotometrically and the preceding kinetic observations were deduced from the relation

$$- d\ell n(\text{opt. density } \lambda_{max} \, 334)_t/dt = \text{const.}[D]/(\text{opt. density } \lambda_{max} \, 334)_0^{1/2} \, .$$

The following mechanism accounts for these results. The unassociated lithium polystyrene is assumed to be the only species involved in the addition, i.e.

$$\sim \overline{S}, Li^+ + D \rightarrow \sim S\overline{D}, Li^+ \, , \quad k \, .$$

However, the bulk of lithium polystyrene, as well as of the products, $\sim S\overline{D}, Li^+$, form homo- or hetero-dimers,

$$2(\sim \overline{S}, Li^+) \rightleftharpoons (\sim \overline{S}, Li^+)_2 \, , \qquad\qquad\qquad ½K_1 \, ,$$

$$2(\sim S\overline{D}, Li^+) \rightleftharpoons (\sim S\overline{D}, Li^+)_2 \, , \qquad\qquad\qquad ½K_2 \, ,$$

and

$$(\text{\wavy} \overline{S}, Li^+) + (\text{\wavy} S\overline{D}, Li^+) \rightleftharpoons (\text{\wavy} \overline{S}, Li^+, \text{\wavy} S\overline{D}, Li^+) \, , \qquad\qquad K_{12} \, .$$

The unassociated species represent only a minute fraction of the dimers. The absorbance of the $- \overline{S}, Li^+$ or $- S\overline{D}, Li^+$ end-groups is assumed to be unaffected by their state of association, whether in the form of a homo- or a hetero-dimer. With these assumptions the solution of the equations describing these equilibria leads to the observed relations, provided that $K_{12} = (K_1 K_2)^{1/2}$. Thus,

$$- d\ell n(\text{opt. density } \lambda_{max} \ 334)/dt = d\ell n(\text{total } [\text{\wavy} \overline{S}, Li^+])_t/dt$$

$$= k \cdot [D]/K_1^{1/2} \ (\textit{initial} \text{ concentration of } \textit{all } \text{\wavy} \overline{S}, Li^+)^{1/2} \, .$$

The interesting feature of this reaction is its "memory". For example, the rate of addition of 1,1-diphenyl ethylene to lithium polystyrene is proportional to the concentration of the remaining living polystyrene. Denoting its concentration at time t by C_L and the rate of addition by R, one finds $R \sim C_L$. However, the higher is the initial concentration of living polystyrene, C_{LO}, the lower the slope of the line. Hence, the rates of two reactions observed at that stage of the process when C_L has some specified value are different whenever they ensued at different initial concentration of lithium polystyrene. This is shown in Fig. 61. The peculiarity discussed here arises from the retarding effect of the product, $\text{\wavy} SD^-, Li^+$. The higher the initial concentration of living polystyrene, $C_{L,0}$, the more $\text{\wavy} SD^-, Li^+$ is formed when the residual concentration of the substrate reaches a specified value. Since $\text{\wavy} SD^-, Li^+$ binds the monomeric $\text{\wavy} S^-, Li^+$ through mixed dimerization, it reduces its concentration in the reacting mixture and slows down the reaction.

Identical behavior, accounted for by the same mechanism, was observed in the addition of styrene to $\text{\wavy} S\overline{D}, Li^{+\ 403)}$, viz.

$$- d\ell n(\text{opt. density } \lambda_{max} \ 460)/dt = k'[S]/K_2^{1/2}(\textit{initial} \text{ concentration of } \text{\wavy} S\overline{D}, Li^+)^{1/2} \, ,$$

where S denotes a molecule of styrene and k' refers to the reaction,

$$\text{\wavy} S\overline{D}, Li^+ + S \rightarrow \text{\wavy} SD\overline{S}, Li^+ \, , \quad k' \, .$$

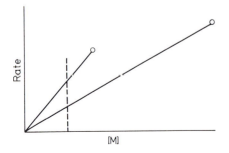

Fig. 61. Plots of rate of addition of 1,1-diphenyl ethylene to lithium polystyrene in benzene vs. total concentration of the unreacted polystyrene. The circles correspond to the initial concentrations of lithium polystyrene. Note the different rates of reactions for the *same momentary concentration of the unreacted lithium polystyrene*, manifesting a kind of "memory"

The importance of mixed dimerization is shown by the results of copolymerization of styrene and p-methyl styrene initiated by lithium alkyls in benzene. O'Driscoll and Patsiga[408], who studied this reaction, reported a linear decrease of logarithms of the concentration of each monomer with time, a relation depicted by Fig. 62 a. The previously discussed scheme accounts for these results[409], namely

$$2(\wwbar{S}, Li^+) \rightleftharpoons (\wwbar{S}, Li^+)_2 , \qquad\qquad 1/2\,K_1 ,$$
$$\quad u \qquad\qquad u_2$$

$$2(\ww pMe\bar{S}, Li^+) \rightleftharpoons (\ww pMe\bar{S}, Li^+)_2 , \qquad\qquad 1/2\,K_2 ,$$
$$\quad v \qquad\qquad v_2$$

$$(\ww\bar{S}, Li^+) + (\ww pMe\bar{S}, Li^+) \rightleftharpoons (\ww\bar{S}, Li^+, \ww pMe\bar{S}, Li^+) , \qquad\qquad K_{12} ,$$
$$\quad u \qquad\qquad v \qquad\qquad uv$$

with the assumed four kinetic steps

$$u + S \rightarrow u , \quad k_{11} ; \quad v + S \rightarrow u , \quad k_{21} ,$$
$$u + pMeS \rightarrow v , \quad k_{12} ; \quad v + pMeS \rightarrow v , \quad k_{22} .$$

The meaning of the symbols is self-evident. The slopes of the logarithmic plots, denoted by λ_1 and λ_2, are given therefore by

$$k_{11}u + k_{21}v = \lambda_1 \quad \text{and} \quad k_{12}u + k_{22}v = \lambda_2 ,$$

and since the total concentration of living polymers is constant and equals I_0, due to absence of termination, an additional equational equation has to be fulfilled, viz.

$$K_1u^2 + 2K_{12}uv + K_2v^2 = I_0$$

(I_0 denotes the concentration of the initiator quantitatively converted into living polymers). Since u and v are two *different* variables, *not related* by the conventional equation describing the stationary state of radical co-polymerization, i.e. $k_{21}v[S] \neq k_{12}u[pMeS]$, the above three equations have to be identical, i.e. they have to represent only *one* relation of u to v. This is possible only if

$$k_{11}/k_{12} = k_{21}/k_{22} = K_1^{1/2}/K_2^{1/2} \quad \text{and} \quad K_{12} = (K_1K_2)^{1/2}$$

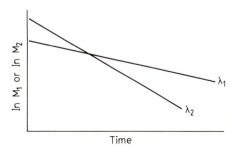

Fig. 62 a. Plots of ℓn of concentration of styrene or p-methyl styrene vs. time for a co-polymerization of those two monomers initiated by butyl lithium in benzene. Both monomers disappear in a pseudo-first order fashion. The pseudo-first order constants are λ_1 and λ_2

The last condition, $K_{12} = (K_1 K_2)^{1/2}$, need not be assumed as had to be done in the previous treatment, it is deduced as a consequence of the experimentally observed linear relations of $\ell n[S]$ or $\ell n[pMeS]$ with time. The proposed mechanism predicts the proportionality of λ_1 and λ_2 with $I_0^{1/2}$, i.e.

$$\lambda_1 = \gamma_1 I_0^{1/2} \quad \text{and} \quad \lambda_2 = \gamma_2 I_0^{1/2}, \quad \text{with } \gamma_1 \text{ and } \gamma_2 \text{ constant .}$$

This relation, not noted by the authors, is borne out on inspecting their data. Its deliberate verification would be desired.

The interesting co-polymerization of butadiene and styrene initiated in hydrocarbon solvents by lithium alkyls was first described by Korotkov[410]. After completion of the initiation the reaction proceeds slowly and virtually only butadiene polymerizes. When this monomer is exhausted, the reaction speeds up and styrene is incorporated then into the polymers. This is clearly shown in Fig. 62 b. Spectroscopic observations indicate that hardly any polystyryl lithium is present in the reacting system while the butadiene monomer is still available; but that ion is formed and becomes abundant when the concentration of butadiene becomes negligible.

This unusual behavior was accounted for by O'Driscoll and Kuntz[411], who postulated an extremely fast addition of butadiene to lithium polystyryl but slow addition of styrene to lithium polybutadienyl. Such a relation prevents the formation of any significant proportion of lithium polystyryl in the polymerizing system as long as butadiene monomer is available. Since the homopolymerization of butadiene is relatively slow, whereas that of polystyrene is rather fast, the polymerization speeds up as butadiene disappears.

These ideas were confirmed by direct studies of kinetics of addition of butadiene to lithium polystyrene and styrene to lithium polybutadiene[412, 413]. The first reaction was too fast to be followed. An estimate, based on the assumption of its internal first order kinetics in living polystyrene, gives the bimolecular constant of butadiene addition to lithium polystyrene in benzene as ~ 1.3 $M^{-1}s^{-1}$ at 29 °C, *independent* of the initial concentration of lithium polystyrene which was varied from $1 - 4 \cdot 10^{-3}$ $M^{[412]}$. This is a

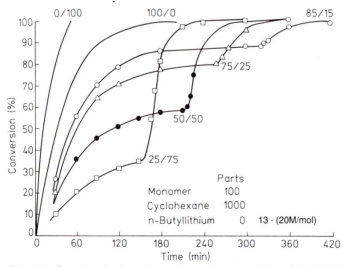

Fig. 62 b. Copolymerisation of butadiene and styrene initiated by BuLi in cyclo-hexane at 50 °C

strange result, implying a direct addition of butadiene to the dimerc lithium polystyrene, contrary to the homopolymerization*.

The addition of styrene to lithium polybutadiene was shown to be by three orders of magnitude slower than the addition of butadiene to lithium polystyrene. Morton and Ells found this reaction to obey the internal first order kinetics in lithium polybutadiene, but the observed first order rate constant decreases linearly with the square root of the initial concentration of lithium polybutadiene.

The results of Morton and Ells and of Johnson and Worsfold led to the following constants:

Co-Polymerization of Butadiene (B) and Styrene (S) in $M^{-1}s^{-1}$ Units

	Morton and Ells[412] Benzene, 29 °C	Johnson and Worsfold[413] Cyclohexane, 40 °C
$10^2 \cdot k_{BB}$	1–2 (?)	10
$10^3 \cdot k_{BS}$	3–7	~ 4
k_{SB}	~ 1	Very Fast
$10^2 \cdot k_{SS}$	5–10	~ 5

In spite of the differences in solvents and temperatures, the results are similar and reveal the same pattern, namely, $k_{SB} \gg k_{BS}$, $k_{BB} \gg k_{BS}$ and $k_{SS} > k_{BB}$, accounting for the unconventional behavior of this polymerization.

In his pioneering work, Korotkov proposed an unusual explanation for the butadiene – styrene co-polymerization. Butadiene was assumed to coordinate rapidly to the Li^+ cation, forming a strong complex. This complex slowly and intra-molecularly yields an $n + 1$-mer out of n-mer, the incorporation of butadiene being rapidly followed by the complexation with another butadiene molecule. The complexation prevents the addition of the intrinsically more reactive styrene.

This hypothesis was disproved by Worsfold[413, 415] who investigated the effect of butadiene on the rate of addition of styrene to lithium polystyrene. Only qualitative results were reported; the rapidity of the butadiene addition prevented quantitative investigation. However, a closely similar system, isoprene – styrene, was more amenable to a quantitative study[415] because the addition of isoprene to lithium polystyrene is somewhat slower. The general pattern of co-polymerization is the same as in the butadiene – styrene system. The effect of small amounts of isoprene on the rate of styrene addition was investigated and the retarding effect demanded by the Korotkov mechanism was not observed**.

Addition of lithium polystyrene to bis-1,1-diphenyl ethylene derivatives reveals some interesting effects caused by the homo- and hetero-aggregation[414]. The reaction proceeded in benzene, and the following substrates, prepared by Tung et al.[416], were investigated:

---

* The addition of styrene to lithium polystyrene follows a square root dependence of the rate on lithium polystyryl concentration
** Concentration of living polystyrene in this system was determined spectrophotometrically

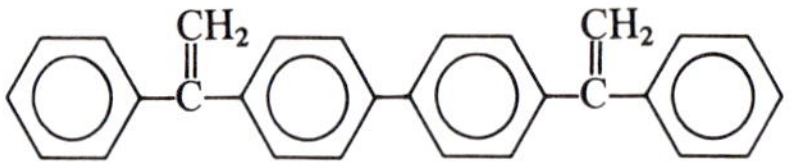

= D–D

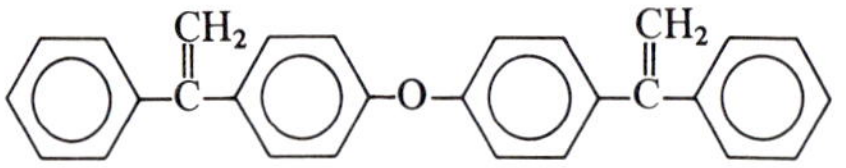

= DOD

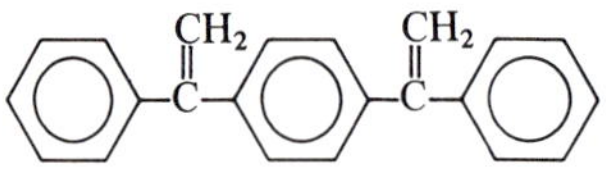

= B

Only one of the C=C bonds reacts when the bis-compound is in excess. Kinetics of that addition is governed by the homo- and hetero-dimerizations, as outlined by Laita and Szwarc[402]; the condition $K_{12} = (K_1K_2)^{1/2}$ is fulfilled for the DOD system, while $K_{12} > (K_1K_2)^{1/2}$ for the D–D and B system[414].

Two consecutive additions, to the first and then to the second C=C bonds, are observed when lithium polystyrene is in excess:

$$D–D + Li^+, \overline{S} \sim \rightarrow D–\overline{D} \cdot S \sim \qquad (a)$$
$$\qquad\quad Li^+$$

$$\sim S \cdot \overline{D}–D + Li^+, \overline{S} \sim \rightarrow \sim S \cdot \overline{D}–\overline{D} \cdot S \qquad (b) ,$$
$$\qquad\quad Li^+ \qquad\qquad\qquad\qquad\qquad\qquad Li^+ Li^+$$

where $\sim \overline{S}, Li^+$ denotes an unassociated lithium polystyrene. Both reactions were simultaneously monitored by a spectrophotometric technique, reaction (a) being considerably faster than (b). In these experiments polystyryl was in a large excess and therefore the ultimate product of reaction (a) had to be a hetero-adduct

$$(D–\overline{D} \cdot S \sim, Li^+, S \sim) .$$
$$\quad Li^+$$

Its conversion into the product of reaction (b) could arise, therefore, from an intra-molecular process, namely,

$$(D–\overline{D} \cdot S \sim , Li^+, \overline{S} \sim) \rightarrow \sim S \cdot \overline{D}–\overline{D} \cdot S ,$$
$$\quad Li^+ \qquad\qquad\qquad\qquad\qquad Li^+ Li^+$$

and the resulting dianion would be subsequently stabilized by complexation with the still available lithium polystyrene yielding the bis-hetero-dimer

$$\left(\begin{array}{c} \sim S \cdot \overline{D}–\overline{D} \cdot S \sim \\ Li^+ \ Li^+ \\ \sim \overline{S}, Li^+, Li^+, \overline{S} \sim \end{array} \right)$$

Indeed, the conversion obeys first order kinetics. However, the first order rate constant was found to increase proportionally with square-root of lithium polystyrene concentration, implying a bimolecular process:

$$(\wedge\!\!\wedge S \cdot \overline{D} - D, Li^+, \overline{S} \wedge\!\!\wedge) + Li^+, \overline{S} \wedge\!\!\wedge \rightarrow (\wedge\!\!\wedge S \cdot \overline{D} - \overline{D} \cdot S \wedge\!\!\wedge, Li^+, \overline{S} \wedge\!\!\wedge)$$
$$\quad\quad Li^+ \quad\quad\quad\quad\quad\quad\quad\quad\quad\quad\quad Li^+ \; Li^+$$

followed by further aggregation of the primary product with $\wedge\!\!\wedge \overline{S}, Li^+$. It would be interesting to investigate the conversion (b) in a $1:2$ mixture of D–D and lithium polystyrene when the presumably formed adduct

$$(\wedge\!\!\wedge S \cdot \overline{D}\text{–}D, \; Li^+, \; \overline{S} \wedge\!\!\wedge)$$
$$\quad Li^+$$

is present virtually in the absence of any additional lithium polystyryl. Unfortunately, this was not done, although such a conversion could easily be observed due to the appearance of a strong absorbance of the dianion ($\lambda_{max} = 598$ nm, $\varepsilon = 14 \cdot 10^4$).

That kind of a problem seems to arise in the addition of sec-butyl lithium to di-isopropenyl benzene. The reaction proceeding in benzene was investigated by Rempp et al.[407]. The investigators intended to develop a dilithio initiator soluble in hydrocarbon. Therefore, they prepared a stoichiometric mixture of the reagents (2:1). Protonation of such a mixture, presumably after cessation of the reaction, yielded butane and the mono-adduct,

implying the hetero-dimeric structure of the adduct formed in the reaction,

However, on addition of styrene or butadiene, a bifunctional polymer was produced*. It seems that the second step of the addition is prevented by the lack of stabilization of the di-adduct. On the other hand, the reaction

---

* This conclusion was partially questioned in a later paper[504]

would lead to a crowded and sterically unfavorable bis-dimer. However, the reaction of the solvated BuLi, attached to a polymer formed after the monomer addition, with the second C=C bond of another polymer, i.e.

$$(\text{\textasciitilde\textasciitilde\textasciitilde}\bar{S}, \text{Li}^+, \text{sec-BuLi}) + CH_2{:}C \underset{\text{(ring)}}{\overset{CH_3}{\Big|}} \quad C \overset{CH_3}{\overset{\Big|}{\underset{S\text{\textasciitilde\textasciitilde}}{\diagup}}} {}^{CH_2Bu} \longrightarrow$$

$$\left(\text{\textasciitilde\textasciitilde}\bar{S}, \text{Li}^+; \bar{C} \underset{CH_3 \quad CH_2Bu}{\overset{Li^+}{\diagdown \diagup}} C \cdot S\text{\textasciitilde\textasciitilde} \underset{CH_3 \quad CH_2Bu}{\diagup} \right)$$

would lead to a feasible and less crowded dimer. The latter reacting with the still available monomer yields then a bifunctional living polymer.

The kinetics of sec-BuLi addition to m-di-isopropenyl benzene was investigated[417]. However, a large excess of sec-BuLi was used in those studies, eliminating the problems of the unfavorable dimerization. Not surprisingly, both C=C bonds reacted under those conditions and were found to be virtually iso-reactive, i.e. the addition of sec-BuLi to one of them did not effect the reactivity of the other.

Another attempt at preparing bifunctional polymers in hydrocarbon solvents with lithium counter-ions was reported by Sigwalt et al.[440]. The following compounds were prepared

$$CH_2 : C(Ph) \cdot (CH_2)_n \cdot C(Ph) : CH_2$$

and

$$CH_2{:}C \overset{CH_3}{\overset{\Big|}{\diagup}} \text{\Large\textcircled{}} (CH_2)_n \text{\Large\textcircled{}} \overset{CH_3}{\overset{\Big|}{\diagdown}} C{:}CH_2$$

They react with sec- or t-BuLi and are converted into insoluble dilithiated derivatives which become solubilized on addition of monomers and yield bifunctional polymers and A.B.A. block polymers. Useful dilithio-initiators were reported by Tung[475].

In all attempts of that kind a difficulty arises from the possible reaction of partially lithiated bifunctional molecule with a fully lithiated one leading to trifunctional dimers. Further reaction might lead to tetrafunctional units, etc.

The aggregation of lithium salts of living polymers in hydrocarbons is disrupted by their coordination with added solvating agents like ethers or amines. As a result the propagation is accelerated* and its stereochemistry is greatly modified. The complexity

---

* Ethers and amines accelerate initiation to a much greater degree than propagation. Therefore, their effect on the rate of overall polymerization, initiation and propagation, is more pronounced than would be expected from consideration of propagation alone

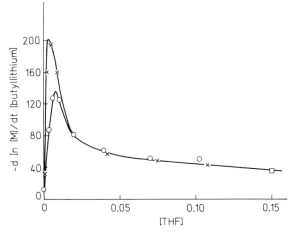

Fig. 63. The standardized rates of styrene polymerization, $d\ell n[\text{styrene}]/dt$ [butyl lithium], at different concentrations of added THF. The reaction proceeds in benzene at ambient temperature

of these interactions is illustrated by Fig. 63, showing the spectacular effects exerted by small amounts of THF on the rate of propagation of lithium polystyrene in benzene[418]. The rate sharply increases and reaches a maximum when the concentration of the ether slightly exceeds the concentration of the living polymer. On further increase of the ether concentration, the rate decreases and eventually it levels off at the THF concentration of ~ 0.1 M. Moreover, at $[\text{THF}] < 10^{-3}$ M, the rate of propagation is proportional to square root of the living polymer concentration, like in the absence of THF. However, it becomes first order in lithium polystyrene at $[\text{THF}] > 0.1$ M. The authors concluded that the more reactive $1:1$ complex of THF and the uncomplexed lithium polystyrene are converted into a less reactive $2:1$ complex as the concentration of THF increases. The bimolecular propagation constant of the dietherate is ~ 0.5 $M^{-1}s^{-1}$ at ambient temperature, i.e. at 10^{-3} M concentration of lithium polystyrene the propagation is 4–5 times faster in the presence of a large excess of the ether than in its absence.

A somewhat similar phenomenon is observed in polymerization of orthomethoxy styrene initiated by butyl lithium in toluene[431]. The initiation is virtually instantaneous and the rate of propagation much faster than of styrene under the same conditions since the intra-molecular complexation with the methoxy group greatly facilitates the dissociation of the dimers. The plot of ℓn of the rate of propagation vs. ℓn of living polymers concentrations is not linear. The kinetic order of the reaction, given by the slope of that plot, increases from 0.5 at $2 \cdot 10^{-2}$ M concentration of living polymers to 0.67 at its $4 \cdot 10^{-4}$ M concentration. The dissociation constant of dimeric into monomeric polymers is determined from the curvature of the plot, namely $K_{\text{diss}} \sim 10^{-3}$ M at ambient temperature. On that basis, the propagation constant of the unassociated polymers is ~ 1 $M^{-1}s^{-1}$ at 21 °C.

A measurable degree of dissociation of other alkali salts of living polystyrene in benzene was reported by Roovers and Bywater[432]. For the sodium salt the dissociation was too low to be determined, while the dissociation of the cesium salt was virtually quantitative in the range 10^{-3} to 10^{-5} M. The potassium salt showed an intermediate behavior allowing for the determination of the respective K_{diss} as $6 \cdot 10^{-4}$ M. The absolute propagation constants of the monomeric living ion-pairs were reported to be 47 $M^{-1}s^{-1}$, 24 $M^{-1}s^{-1}$ and 18 $M^{-1}s^{-1}$ for the K^+, Rb^+ and Cs^+ pairs, respectively, at 25 °C, implying an important contribution of solvation by benzene in the transition state of propagation.

Spectrophotometric determination of the dissociation of an aggregated lithium poly-isoprenyl in benzene and in n-octane led to the values of the dissociation constant given below[433]:

T °C	$10^5 K_{diss}$/M in benzene	$10^7 K_{diss}$/M in n-octane
11.5	0.37	~1.1
20.5	0.62	~2.5
30	~1.0	5–5.5
40	~1.5	~9
58	–	28

On this basis the enthalpy of dissociation in benzene is ~9 kcal/mol and in n-octane ~12 kcal/mol in agreement with the estimated upper limit of 14–15 kcal/mol. The impossibly high value of 37 kcal/mol was shown to be without any foundation[171].

The problem of the nature of the dissociation process needs clarification. The authors believe that lithium polyisoprenyl is aggregated into tetramer and dissociates into dimers, and the extremely low dissociation of the latter yields the propagating monomers.

IV.14. Anionic Polymerization of Oxiranes and Thiiranes

Extensive studies of anionic polymerization of ethylene oxide and its derivatives were reported during the last forty years. Those reactions initiated by coordination catalysts, like metallic oxides, have been reviewed briefly in the Introduction. Early investigators of polymerization initiated by alkoxides dealt with systems involving alcohols which solubilized the scarcely soluble alcoholates. Since alcohols act as chain-transfer agents, living polymers are not formed in those systems. One could expect to overcome the difficulties caused by lack of solubility by performing the polymerization in aprotic but powerfully solvating media since termination and chain-transfer might be avoided in those solvents. With this goal in mind, Figueruello and Worsfold[51] investigated poly-merization of ethylene oxide initiated by the sodium or potassium salts of mono-methyl ether of diglyme, $CH_3OC_2H_4OC_2H_4O^-$, Alk^+, in hexamethyl phosphorictriamide, HMPA. However, the polymerization turned out to be complex and the results were not readily interpreted.

In this aprotic solvent strongly interacting with cations, termination and chain-trans-fer were avoided as evident by the relation:

Number average degree of polymerization = [polymerized monomer]/[initiator],

reliably established for this system. Kinetics of propagation of the sodium salt revealed its first order character in respect to the monomer. However, the rate of propagation was found to be independent of the concentration of living polymers varied from $7 \cdot 10^{-5}$ M to about 10^{-2} M. The first order propagation constant, $-d\ell n[M]/dt$, was reported as $\sim 3 \cdot 10^{-4} \, s^{-1}$ at 40 °C.

The kinetic findings were supplemented by the results derived from studies of conductance and viscosity of polymerized solutions. The equivalent conductance was low and virtually constant in the range $\sim 4 \cdot 10^{-2}$ M down to $\sim 10^{-3}$ M, although its value increased 5-fold on further dilution to $6 \cdot 10^{-5}$ M. In spite of the relatively high dielectric constant of HMPA, 26 at 40 °C, alkoxides behave in that solvent like weak electrolytes in contrast to their behavior in methanol. In the latter solvent the dissociation of, say sodium methoxide, is virtually quantitative at 10^{-3} M concentration[434], an observation accounted for by a strong hydrogen bonding of the methoxide ion to methanol. On the other hand, sodium tetraphenylboride behaves in HMPA as a strong electrolyte, and its addition to polymerizing solutions retards the propagation. This observation implies that the free $\sim\sim CH_2CH_2O^-$ anions propagate faster than their ion-pairs.

The viscosity of living polymers solutions decreased on termination of the active end-groups implying their association into some unspecified aggregates. In fact, the authors suggest that the aggregation is responsible for the complex kinetics of propagation. However, the low degree of direct dissociation of alkoxides pairs, hindered by the lack of stabilization of alkoxides anions, does not preclude the formation of ions through the reaction,

$$2\,RO^-, Na^+ \rightleftharpoons RO^-, Na^+, O^-R + Na^+ \,.$$

Such a process could account for the independence of polymerization rate on the concentration of living polymers, provided that the RO^- anions, arising from the buffered dissociation $RO^-, Na^+ \rightleftharpoons RO^- + Na^+$, are the main contributors to the propagation. The dissociation of the $(RO^-, Na^+)_n$ aggregates, taking place at higher dilution, would account then for the increase in the equivalent conductance without affecting the rate of propagation.

The behavior of the potassium salt is more conventional. In that system, propagation is again first order in monomer, but its first order propagation constant, $k_u = - d\ell n[M]/dt$, increases with increasing concentration of living polymers, e.g. being $0.5 \cdot 10^{-4} \, s^{-1}$ at 10^{-4} M concentration and $\sim 3 \cdot 10^{-4} \, s^{-1}$ at $13 \cdot 10^{-4}$ M. However, a further increase of living polymers concentration only slightly affects k_u, e.g. at concentration of $140 \cdot 10^{-4}$ M $k_u = 5 \cdot 10^{-4} \, s^{-1}$.

It seems that the aggregation of ion-pairs is less pronounced in the potassium system than in the sodium, while their direct dissociation into free ions, as well as their capacity to form triple ions, is enhanced.

A simple behavior is observed in the polymerization of cesium salt[423]. The propagation is again first order in respect to monomer and the observed bimolecular propagation constant, $k_p = (- d\ell n[M]/dt)/[\text{living polymers}] = k_u/C_L$, is found to be a linear function of $1/C_L^{1/2}$ in the absence of borides, while it is a linear function of $1/[Cs^+]$ in the presence of cesium tetraphenyl boride. These relations are illustrated by Fig. 64. From the slopes of the lines shown in that figure, the propagation constant of the free anions was calculated as $k_- = 22 \, M^{-1}s^{-1}$ and the dissociation constant of the ion-pairs as 10^{-5} M, both at 40 °C. The latter value was confirmed by conductance study. The intercept of the lines shown in Fig. 64 gives the propagation constant of ion-pairs $k_\pm = 0.2 \, M^{-1}s^{-1}$.

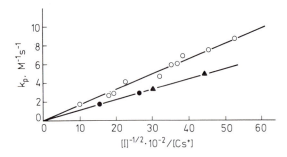

Fig. 64. Plots of bimolecular propagation constant, k_p, of ethylene oxide polymerization in THF with Cs^+ counter-ions vs. reciprocal of square-root concentration of initiator, $1/I^{1/2}$ (*open circles*) or reciprocal of Cs^+ concentration $1/[Cs^+]$. The concentration of cesium varied by addition of $CsBPh_4$

Much work was done on polymerization of ethylene, propylene, and iso-butene oxides in dimethyl sulphoxide, CH_3SOCH_3. Blanchard et al.[424] studied the polymerization of iso-butylene and propylene oxides initiated by potassium *t*-butoxide and found its rate to be first order in the monomer but nearly second order in the initiator. The linear log-log plots of rate vs. the initiator concentration had slopes 1.8 (for iso-butylene oxide) and 1.7 (for propylene oxide). The former polymerization was complicated by phase separation – the polymer being insoluble in the solvent, whereas some chain-transfer to the solvent was observed in the propylene oxide system. Again, high order in the initiator, namely 1.9, was reported for the polymerization of ethylene oxide initiated by $C_2H_5O \cdot C_2H_4 \cdot OC_2H_4 \cdot O^-$, Cs^+ in dimethyl sulphoxide[436], as shown by Fig. 65. On the other hand, first order dependence on the alkoxides' concentration was claimed by Bawn et al.[435], and by Price and Carmelite[438]. An interesting attempt to reconcile these contradictory findings was reported by Figueruello[422]. As revealed by conductomeric work of Steiner[437], complex equilibria are established in the alkoxide-dimethyl sulphoxide systems, namely

$$RO^- + CH_3SO \cdot CH_3 \rightleftharpoons ROH + CH_3SO \cdot CH_2^- \tag{a}$$

$$RO^- + ROH \rightleftharpoons RO^-(HOR) \tag{b}$$

$$RO^-(HOR) + ROH \rightleftharpoons RO^-(HOR)_2 \tag{c}$$

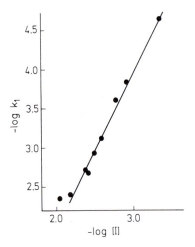

Fig. 65. Plot of log of bimolecular propagation constant of ethylene oxide polymerization in dimethyl sulphoxide vs. log of concentration of the initiator, $C_2H_5OC_2H_4OC_2H_4O^-$, Cs^+. Slope 1.9

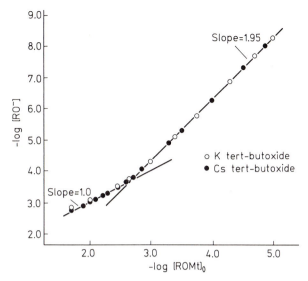

Fig. 66. Log of propagation constant of ethylene oxide polymerization in dimethyl sulphoxide vs. log of initiator concentration. The *solid lines* are computed by taking into account the equilibria investigated by Steiner. Note the 2 regions – at higher concentrations of the initiator second order dependence is observed, whereas a first order dependence is found for higher dilution

as well as the equilibrium of pairing anions with cations. It seems that the free alcoholate anions are the most reactive species contributing to the propagation. Equilibrium (a) is responsible for chain-transfer leading to the formation of alcohols. Using the equilibrium data reported by Steiner, Figueruello calculated the concentration of RO$^-$ ions as a function of the added RO$^-$, Cat$^+$ as shown in Fig. 66. The results reveal two regions of alcoholate concentrations; at its higher dilution the concentration of [RO$^-$] depends on 1.95 power of the initiator concentration, whereas it is proportional to its first power at its higher concentration. Since the rate of propagation is assumed to be proportional to the concentration of the free alkoxide, the controversial claims seem to be reconciled.

Kinetics of polymerization of ethylene oxide in tetrahydrofuran was reported by Kazanskii et al.[52]. The study was complicated by strong association of the poly-alkoxides. In fact, its extent is dramatically revealed by the gelation observed on addition of ethylene oxide to THF solutions of living polystyrene with two active end-groups[444]. In Kazanskii's study the association was manifested by the low fractional kinetic order of propagation, as illustrated in Fig. 67. The rate of propagation was shown to be ¼ order in concentration of the sodium alkoxides and about ⅓ order for the potassium and cesium salts. On this basis the propagation constants of the unassociated alkoxides was calculated to be 0.94 M^{-1}s^{-1} for the potassium salt and 3.5 M^{-1}s^{-1} for that of cesium. The contribution of the free alkoxide ions was discarded in view of the extremely low dissociation constant of alkoxide ion-pairs in THF, viz. $7 \cdot 10^{-10}$ M for Cs$^+$ and $8 \cdot 10^{-11}$ for K$^+$ salts[52(b), 445]. Determination of ΔH and ΔS of the association was reported; however, the results should be considered as tentative.

The association is avoided by complexing the cations with suitable kryptates. Polymerization of ethylene oxide initiated by potassium carbazyl in THF in the presence of

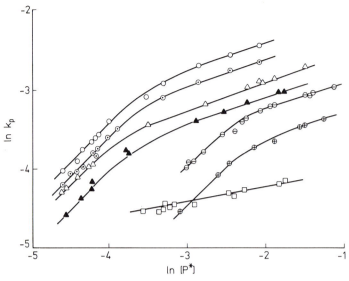

Fig. 67. Logarithmic plots of propagation constant of epoxides polymerization vs. concentration of growing polymers

2,2,2-kryptate proceeds smoothly, termination and transfer being prevented. The association of the alkoxide was negligible at their concentrations lower than 6.10^{-4} $M^{54)}$. The dissociation constant of the kryptated potassium salt was determined from the dependence of the rate on $1/C_L^{1/2}$ in the absence of added tetraphenyl boride and on $1/[2,2,2,K^+]$ in its presence, viz. $K_{diss} = 3 \cdot 10^{-7}$ M. The dissociation constant of kryptated potassium tetraphenyl boride was determined from its conductance. A plot of the observed bimolecular rate constant, k_p, vs. f,fraction of dissociated ion-pairs, is shown in Fig. 68. The $k_\pm$ and k_- were found to be 1.5 $M^{-1}s^{-1}$ and 100 $M^{-1}s^{-1}$ at 20 °C in THF.

These investigations were extended later to the cesium salt, using the spheroidal kryptate denoted by TC as the complexing agent[55]. Its synthesis was reported by Graf and Lehn[442] and its structure is given below:

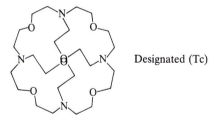   Designated (Tc)

The conductance study of $(Cs^+, TC)BPh_4^-$ in THF led to its dissociation constant, of $1.1 \cdot 10^{-4}$ M at 20 °C. A perfect linear relation was obtained for k_p plotted vs. f,fraction of dissociated ion-pairs, yielding $k_\pm$ 5.6 $M^{-1}s^{-1}$ and $K_{diss} \sim 1 \cdot 10^{-6}$ M at 20 °C. Hence, the kryptated cesium salt is about four times as reactive as the kryptated potassium salt, and the dissociation constant of the former salt is three times larger than of the latter.

Anionic, base catalyzed polymerization of propylene oxide and higher epoxides are complicated by the E2 elimination acting as a chain transfer, e.g.

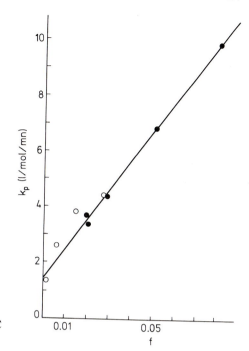

Fig. 68. Plot of propagation constant, k_p, vs. mole fraction of free polyethylene oxide anion in polymerization proceeding in THF at $-30\,°C$ with $K^+(2,2,2)$ counter-ions

$$\text{wwO}^- + CH_3\cdot CH \overset{O}{\underset{\triangle}{-}} CH_2 \xrightarrow{\; k_{tr} \;} \text{wwOH} + CH_2{:}CH\cdot CH_2O^-$$

This reaction limits the maximum degree of polymerization of based catalyzed polymerization; at operable temperature of propylene oxide polymerization, the ratio k_p/k_{tr} ~ 100 [448].

An interesting problem arises in polymerization of propylene oxide. This cyclic monomer possesses an assymetric carbon atom and therefore it exists in two enantiomorphic forms. The question arises whether, under the same conditions, propagation of a pure enantiomer proceeds with the same rate as the propagation of the racemic mixture. Such a study was reported by Price [53, 448]. Polymerization initiated by potassium t-butoxide in dimethyl sulphoxide or in hexamethyl phosphoric-triamide was investigated at several temperatures and the following results were reported:

Solvent	T°C	Propagation Constant in 10^{-4} $M^{-1}s^{-1}$ Units	
		Pure Enantiomer, S	Racemic Mixture
DMSO	25	~ 1.1	~ 2.4
	30	3.6	~ 7.0
	40	~ 4.6	~ 13.0
	55	–	44.0
HMPA	15	1.4	22.6
	25	3–7	6–15
	40	10.0	29.0

Although the experimental scatter of the results is larger than desired, the results conclusively show the preference for some kind of alternation in polymerization of racemic mixture resulting in formation of polymers built from regular sequences of $dd\ell\ell$ units. Indeed, examination of their NMR spectra revealed an equal proportion of isotactic and syndiotactic placements[441], indicating either the formation of long blocks of isotactic and syndiotactic sequences or the proposed kind of alternation. Additional evidence was provided by studies of degradation products (glycols) of poly-propylene oxide and poly-t-butylene oxide prepared from the respective racemic mixtures under conditions described above[428]. Study of inversion of configuration in ethylene oxide polymerization was reported by Matsuzaki and Ito[429].

The preference for the alternation is attributed to chelation of the K^+ ion with the oncoming monomer as well as with the terminal and penultimate units of the polymer. The interesting feature of the proposed stereochemistry is that the incoming configuration is always opposite to that of the penultimate units, i.e. $k_{dd\ell*} \gg k_{ddd*}$ and $k_{d\ell\ell*} \gg k_{d\ell d*}$.

Polymerization of chiral monomers was reviewed recently by Sigwalt[430]. Two situations may arise. The non-chiral initiator may interact randomly with either enantiomer, but its parity introduces a bias in the subsequent addition. For a homosteric bias, i.e. when the addition of one enantiomer favors the subsequent addition of the same kind, the resulting polymer is composed of sequences of blocks of one enantiomer followed by a block of the other. This is a typical example of stereoselectivity. For an antisteric bias, i.e. the addition of, say R enantiomer is favored by the presence of terminal S enantiomer, or *vice versa*, the resulting polymer shows a bias for a simple alteration.

A chiral catalyst associated with the growing end-group exhibits a bias for one of the two enantiomers. In such a case, the resulting polymer is optically active, provided that an optically active catalyst was used in the preparation, while the residual monomer becomes enriched in the other enantiomer. This is an example of stereo-electivity. A racemic mixture of chiral catalysts yields a racemic mixture of polymers, each being enriched in one of the enantiomers.

The first report of anionic polymerization of propylene sulphide yielding living polymers appeared in 1963[425]. The polymerization was initiated by sodium naphthalenide in THF and the product had the degree of polymerization expected for a living system (with two active ends). The detailed mechanism of initiation was discussed previously. Kinetics of propagation at $-30\,°C$ was investigated[109 b]. Living polymers form aggregates at concentrations higher than 10^{-3} M, but at lower dilution the system is simple, being composed of free ions and ion-pairs only. The former species are substantially more reactive than the latter, i.e. $k_\pm \sim 10^{-2}$ $M^{-1}s^{-1}$, while the observed overall propagation constant is about 12 $M^{-1}s^{-1}$ at that temperature. The dissociation constant of the sodium salt in THF was determined[450] at $5 \cdot 10^{-8}$ M at $-40\,°C$ and $0.7 \cdot 10^{-8}$ M at $0\,°C$. The presence of the monomer slightly affects the dielectric constant of the solution, causing some increase of K_{diss}. Using these data, as well as the kinetic data obtained from runs involving sodium tetraphenyl boride, the individual rate constants were determined, viz. $k_\pm = 1 \cdot 10^{-3}$ $M^{-1}s^{-1}$ and $k_- = 1.7$ $M^{-1}s^{-1}$ at $-40\,°C$ and $k_\pm = 0.03$ $M^{-1}s^{-1}$ and $k_- \sim 40$ $M^{-1}s^{-1}$ at $0\,°C$[450]. The activaton energy of the free ions propagation was calculated as ~ 10 kcal/mol, corresponding to the frequency factor of $\sim 2 \cdot 10^9$ $M^{-1}s^{-1}$, while the respective values for the Na^+ ion-pairs were reported as 11 kcal/mol and $2 \cdot 10^7$ $M^{-1}s^{-1}$. It was concluded that only one kind of ion-pairs is present in that system. The extension of those

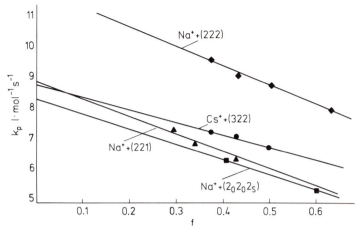

Fig. 69. Plot of propagation constant, k_p, of poly-propylene sulphide polymerization with kryptated cations in THF vs. mole fraction of free ions. Note the increase of k_p at lower f indicating *higher* reactivity of kryptated ion-pairs than of the free ions

studies to other systems was reported, viz. those involving counter-ions like NBu_4^+ [426] or Cs^+ [427], and pertaining to solvents like THP or DME[427]. The propagation constant of cesium poly-propylene sulphide ion-pairs increases from 0.27 in dioxane to 0.5 in THP, to 7–8 in THF or DME, in $M^{-1}s^{-1}$ units at 15 °C.

The most interesting results were obtained in polymerization studies performed with kryptated cations. Kinetics of propylene sulphide polymerization in THF at -30 °C with Na^+, (2,2,2) led to the results shown in Fig. 69. The propagation of these ion-pairs turned out to be *higher* than of the free anions. On the whole propagation constant of ion-pairs increases with the radius of the cation, viz. at -30 °C in THF:

Cation	Na^+	Cs^+	NBu_4^+	$Na^+(2,2,2)$
$k_\pm$ M s	$2.5 \cdot 10^{-3}$	0.23	3.3	11.3
k_- M s	3.8	–	~6 (?)	3.8
r/A	0.95	1.7	4.9	5.5

As could be expected, the complexation facilitates the dissociation of ion-pairs. For example, the dissociation constant of sodium ion-pairs in THF at -30 °C increases one thousand-fold on addition of (2,2,2)kryptate. An interesting study of propylene sulphide polymerization initiated electro-chemically in dimethyl formamide was reported by Mangoli and Daolio[451].

IV.15. Stereochemistry and Tacticity of Polymers

Two kinds of linkages may join vinyl or vinylidene monomers into polymeric chains: the head-to-tail bonds or head-to-head or tail-to-tail linkages. The head-to-tail enchainment usually is preferred in propagation steps for steric or other energetic reasons. Neverthe-less, occasional head-to-head or tail-to-tail bonds were observed and the frequency of their occurrence was determined for some systems, e.g.[469]. The entirely head-to-head-tail-to-tail polymers could be synthesized by special techniques; an example of such a synthesis is described on p. 41.

The head-to-tail linkage of vinyl or unsymmetric vinylidene monomers yields two distinct kinds of bonds referred to as isotactic and syndiotactic, and polymers are consid-ered to be stereoregular when long sequences of their units are linked together by the same kind of linkage. The first fully recognized and well documented synthesis of a stereoregular polymer was reported by Natta and his co-workers in 1955[470], and since then numerous synthesis of other stereoregular polymers have been reported in the literature.

The mechanism of stereospecificity depends on the mode of propagation. Much work in this field was devoted to studies of stereospecificity of heterogeneous polymerizations, especially those induced by the Ziegler-Natta kind of catalysts. Mechanistic studies of stereospecificity of anionic polymerization might be more revealing due to its homogene-ous nature, although often they are complicated by the simultaneous participation of several species in the propagation step.

Early evidence for the stereospecificity was furnished by the X-ray studies of solid polymers. They allow a distinction between the crystallinic isotactic or syndiotactic stereoregular polymers and the amorphous atactic polymers. The scope of such studies was greatly expanded by the introduction of NMR techniques pioneered by Bovey and Tiers[471]. They stressed the magnetic equivalence of the CH_2 protons in each syndiotacti-cally bonded dimeric segment of a high molecular weight vinyl or unsymmetric vinylidene polymer, in contradistinction with their non-equivalence when a dimeric segment is linked through an isotactic bond, even if the rotation around the respective C–C bonds is fast. Hence, the former protons give rise to a single line in the respective NMR spectrum while the presence of the latter protons produces an AB quadruplet. Thus, the NMR technique allows one to determine the proportions of syndiotactic and isotactic diads in a studied polymer, these being referred to as the racemic, r, and mezo, m, diads, respec-tively. The designation of the racemic diads as dd or ll is misleading and inappropriate since in long polymers, when the end-effects could be neglected, their distinction is impossible provided that truly asymmetric centers are not present in the chain.

The NMR spectrum of substituents permits a distinction between three kinds of triads. The isotactic, mm, triad is formed when a monomeric segment bearing the sub-stitutent in question is linked through isotactic bonds to both of its neighboring segments; linkage through two syndiotactic bonds yields a syndiotactic triad, rr, whereas a heterotactic triad, $mr = rm$, results when a monomeric segment is linked to one of its neighbors by an isotactic bond, while a syndiotactic bond links it to the other adjacent unit. For example, the protons of α-methyl groups of poly-methyl methacrylate produce in their NMR spectrum three lines, each corresponding to a different chemical shift, depending upon whether the unit in question is the center of an mm, rr, or mr triad. Their relative intensities give therefore the proportions of isotactic, syndiotactic, and heterotactic triads in a studied polymer.

In a similar fashion, one may determine the proportions of six different tetrads, mmm, $mmr = rmm$, rmr, mrm, $rrm = mrr$, and rrr, e.g.[472], or of ten different pentads, and indeed the newest developments in instrumentation, as well as utilization of [13]C-

NMR etc., make such determinations possible, at least for some systems. For a concise review of this subject, the reader is referred to two monographs by Bovey[381, 477].

The above discussion is concerned with the *configuration* and not the *conformation* of polymer chains. The sharpness of the NMR spectra is optimal when all the chain conformations are equally probable and their interchange is very rapid. Although valuable and interesting information is gained from the complexities caused by conformational problems, they do not pertain to our subject.

Let us consider now the problems of sterospecificity of the addition step in anionic propagation of vinyl or vinylidene monomers. The predominant structure of the growing end-groups is depicted below:*

$$
\begin{array}{ccc}
& \overset{\displaystyle X}{\underset{\displaystyle Y}{-CH_2-\bar{C}\!\!\diagdown}}\, ,\ Cat^+ & \text{or} \quad -CH_2-\overset{\displaystyle Cat^+}{\underset{\displaystyle Y}{C}}-X \\[2mm]
\beta \quad \alpha & & \\[2mm]
\text{(a)} & & \text{(b)}
\end{array}
$$

Case (a) corresponds to free carbanions or to loose ion-pairs with cations readily moving from one to the other side of the trigonal plane. The last carbon atom has no definite parity, provided that its interaction with preceding segments is too weak to affect its behavior and that the rotation around the αC–βC bond is truly free. However, on addition of an oncoming monomer a definite, lasting asymmetry is established around that atom leading to the formation of a mezo or racemic diad. The absence of interaction with the preceding segments makes the formation of either diad equally probable; hence, the resulting polymer chain is perfectly random, i.e. the probabilities of occurrence of a mezo or racemic diads are ½**.

Interaction of the last unit with a penultimate one alters the situation. It affects the αC–βC rotation and therefore on the addition of an oncoming monomer the formation of one diad is more favored than of the other, i.e., the probability of forming a mezo diad, say p, is different from that of a racemic one, 1–p. Let it be stressed, however, that p is independent of the parity of the penultimate α-carbon and therefore the resulting chain is Bernoullian.

Three types of interactions may affect the outcome of the addition: steric factors force the spatial arrangement determined by the bulkiness of substituents[480], polar factors make one configuration advantageous to the other[479], and loose association with a cation in the transition state of the addition affects the proximity of substituents of the last unit and the penultimate one.

In case (b), a definite stereo-structure characterizes the growing end-group. Such a structure, but not its parity, affects the probability of creation of a mezo or racemic diad in a process of formation of a new center, the cation playing an important role in the transition state of the addition. In the absence of interaction between the last and penulti-

---

* There are systems yielding substituents on the β and not on the α carbon, e.g. Ziegler-Natta polymerization of propylene

** The reverse statement need not be true. For example, in a polymer $\cdots rmrm \cdots$ the probability of finding a mezo or racemic diad is again ½, although such a polymer is not a random one

mate units, the resulting polymer chain is Bernoullian with p different from $\frac{1}{2}$, contrasting the result described in case (a). Operation of such interactions makes the probability of a configuration around the newly formed center determined by the nature of the preceding diad, i.e. the resulting chain obeys then the first order Markov statistics requiring two parameters for its description, say r and q. Still more complex relations could be visualized, leading to chains governed by higher order Markov statistics.

Association of growing ends may profoundly affect the mode of addition since it introduces further restrictions on the structure of transition state – a phenomenon similar to that resulting from solvation or coordination of the growing end-group with molecules of solvent or of some complexing agent. Not surprisingly, solvents favoring the association of living polymers often exert a large effect on tacticity of the resulting polymers. Such effects were reported in the literature, e.g.[215, 482]. An interesting situation may be visualized when the stereo-structure of one center of the associate affects the stereochemistry of the other center.

Problems of microtacticity are also encountered in polymerization of dienes. For example, butadiene may be incorporated through 1,2-addition or 1,4-*cis* or 1,4-*trans* addition. Some aspects of that problem were discussed earlier. An observation pertaining to polymerization of lithium poly-butadiene in hydrocarbon solvent may be mentioned here. Makowski and Lynn[483] reported that the proportion of the 1,2-bonds in polybutadiene is abnormally high in the first 10 or more segments compared with the subsequent units. This conclusion was derived from studies of structure of low oligomers and comparing it with that of high-molecular polymers. However, as pointed out by Bywater[478], the oligomers were formed at high concentration of living polymers while the high-molecular polymer was formed at their low concentration. Under conventional conditions, the polymerization proceeds on unassociated polymers since the associates negligibly contribute to the growth at low concentration of living polymers. However, their contribution seems to be appreciable at high concentration of living polymers and apparently the reaction with the associates favors the 1,2-addition, while the 1,4-addition is favored in the propagation involving the unassociated polymers. This explanation of Makowski and Lynn's observation was adequately confirmed by appropriate experiments, shown in Table I of Ref. 478, (see also[481]).

NMR studies of model compounds of lithium polydienes in hydrocarbon solvents reveal that the living end-group has preferentially *trans*-configuration in equilibrium with small amounts of *cis*[405, 485], in contrast to the preference for the *cis*-structure found in polar solvents[486] *. The addition of isoprene to lithium polyisoprene in hydrocarbon solvents is stereospecific and yields the *cis* living end-group which isomerizes to the thermodynamically more stable *trans*-forms. Hence, in fast propagation, i.e. at high monomer concentration, the living *cis*-form is converted into a lasting *cis*-segment, whereas at low monomer concentration the isomerization may take place prior to the addition of the next monomer molecule and the resulting *trans* living end-group is converted into a *trans*-segment. These ideas are clearly depicted by the following scheme:

---

* In this respect, diethyl-ether is an exception. The NMR spectra of hydro-carbon solutions of Li polydienes provide information about the structure of the *aggregates*, whereas the polymerization proceeds with non-aggregated species. The problem arises whether the non-aggregated end-groups have the same structure as the aggregated ones

$$\text{cis}^* + M \xrightarrow{k_{cis}} \text{cis} \cdot \text{cis}^*$$

$$\updownarrow$$

$$\text{trans}^* + M \xrightarrow{k_{trans}} \text{trans} \cdot \text{cis}^* .$$

The propagation constant of $\sim cis^*$ seems to be substantially greater than that of *trans*. A similar scheme applies to polymerization of butadiene; however, the rate of isomerization is much faster and subsequently a relatively large proportion of *trans*-form is incorporated in the polymer. For recent summary see Ref. 487.

The *cis*- or *trans*-mode of opening of the C=C bond of a vinyl monomer, expected to take place in the course of propagation, deserves some comments. The problem is significant only for polymerization involving monomers such as CHD : CXY that exist in two isomeric forms: *cis* and *trans*. A definite and lasting stereo-structure of the CHD group is established on addition of such a monomer to the growing center. However, the configuration around the CXY group still could be altered whenever the growing end-group could rotate, or the counter-ion could move, prior to the addition of a next monomer. Under such conditions it is meaningless to treat the addition as a *cis* or *trans* opening of the C=C bond. Nevertheless, valuable information about the mechanism of propagation was obtained from studies of polymerization of such monomers, see e.g.[362, 477].

A proposed mechanism of polymerization leading to some statistical conclusions about the structure of the resulting polymers gains plausibility if the observed intensities of NMR lines agree with the statistical deduction. However, the agreement does not provide an indisputable proof of the proposed mechanism, whereas the disagreement unquestionably disproves the mechanism. This is a general statement applying to any scientific theory or hypothesis. The hypothesis is discarded whenever it leads to conclusions contradicted by observation, but it is not necessarily correct even when the agreement is perfect.

Some of the relations between the observed intensities of NMR lines are fulfilled, whichever mechanism governs the polymerization[484]. For example,

$$(m) + (r) = 1 \; ; \quad (mm) + (mr) + (rr) = 1$$

$$(m) = (mm) + \tfrac{1}{2}(mr) \; ; \quad (r) = (rr) + \tfrac{1}{2}(mr)$$

$$(mm) = (mmm) + \tfrac{1}{2}(mmr) \; ; \quad (rr) = (rrr) + \tfrac{1}{2}(mrr)$$

$$(mr) = (mmr) + 2\,(rmr) + 2\,(mrm) + (mrr) \; , \text{ etc.}$$

(m), (r), (mm), (mr), etc., denotes the fraction of various diads, triads, etc. in the investigated polymer. Failure of these relations implies an incorrect assignment of the lines. On the other hand, relations such as

$$(mm) = (m)^2 \; ; \quad (mr) = 2\,(m)(r) = 2p(1-p) \quad p = (m)$$

$$(rr) = (r)^2 = (1-p)^2$$

have to be fulfilled for any Bernoullian chain; their failure proves that the investigated chain is *not* Bernoullian. A useful parameter of $\delta = 2\,(m)(r)/(mr)$, referred to as the persistence ratio, was introduced by Coleman and Fox[484]. For any Bernoullian chain, $\delta = 1$, whereas for first order Markov chains $1/\delta = P_{m/r} + P_{r/m}$, where $P_{m/r}$ and $P_{r/m}$ are the probabilities for diad m to yield diad r and for diad r to yield diad m. In other words, $P_{m/r} = (mr)/(\{(mm) + (mr)\}$ and $P_{r/m} = (rm)/\{(rr) + (rm)\}$. As is easily shown, $P_{m/r} + P_{r/m}$ $= 1$ for Bernoullian chains.

For highly isotactic polymers, $P_{m/r} \ll \frac{1}{2}$ while $P_{r/m}$ is large*, $\gg \frac{1}{2}$ but $\leqslant 1$, whereas for highly syndiotactic polymers, $P_{m/r}$ is large while $P_{r/m} \ll \frac{1}{2}$. When both $P_{r/m}$ and $P_{m/r}$ are $\ll \frac{1}{2}$, a block polymer, $mm \cdots mrr \cdots rm \cdots$, is formed, while "alternative-like" polymers (heterotactic), $\cdots mrmrmr \cdots$, are formed when $P_{r/m}$ and $P_{m/r}$ are both close to unity. Note, for block polymers $\delta \gg 1$ and for "alternative-like", $\delta \approx 0.5$.

Most general relations fulfilled by Markov chains of any order have been reported in the literature. Their failure implies a non-Markov chain behavior. Such polymers are formed when two or more interchangeable species participate in the growth, e.g. free ions and ion-pairs, two types of ion-pairs, associated and unassociated living polymers etc.[484, 485]. The structure of the resulting polymer depends then on rate of interchange as well as on the kinetics of propagation and the stochastic characteristic of each species. In a special case of two species, one yielding a pure isotactic sequence while the other producing a pure syndiotactic sequence, the resulting polymer obeys a first order Markov statistic, i.e. it behaves as if the growth were affected by a penultimate unit.

---

* $P_{m/r}$ represents the probability of an error in isotactic polymer, e.g., $\cdots mm \cdots mrmm \cdots$. Two consecutive errors lead to an optically active uniform isotactic polymer, e.g. $\cdots dd \cdots d\ell dd \cdots$, whereas a single error yields isotactic optical blocks, $\cdots dd \cdots d\ell \ell \cdots \ell d \cdots$

V. Concluding Remarks

In spite of its length, this review does not exhaust the subject of anionic polymerization. Some important topics were omitted or only cursorily mentioned. For example, anionic co-polymerization was not discussed comprehensively, although remarks concerning some co-polymerizing systems are found in pp. 145–147. In fact, studies of living polymers permit a direct determination of the cross-over constants denoted usually by k_{12} and k_{21}. In radical co-polymerization, the reactivity ratios, r_1 and r_2, are determined from the functional dependence of the composition of a co-polymer on the composition of the feed. These constants, in conjunction with the absolute homo-propagation constants, lead to the desired results. For living systems, the recourse to reactivity ratios is superfluous. The values of k_{12} or k_{21} could be determined by observing the rates of spectral changes resulting from addition of monomer B to living homo-polymer of A as exemplified by studies of Worsfold and of Morton described on pp. 146–147. This approach was utilized by Van Beylen[499], who investigated the effect of substituents on the cross-over rate constant. Alternatively, one could use living homo-polymer A as an initiator of homo-propagation of B and extrapolate the kinetic results to 0% of conversion[316]. This approach, useful when the cross-over constant is greatly different from the homo-propagation constant of B, was adopted by Shima et al.[500].

Occasionally, unusual effects are observed. For example, addition of styrene to living poly-vinyl-naphthalene results in complexing of the first polystyryl-anion to the preceding naphthalene moiety[501]. Consequently, three constants describe the process: the rate constant of addition of the first styrene molecule, the very low rate constant of addition of the second styrene molecule, and a large rate constant of homo-propagation ensuing from addition of the third, forth, etc., styrene molecules.

Living polymer technique permits also the preparation of active polymers possessing a desired penultimate unit. Thus, the effect of the penultimate unit on the rate constant of propagation could be investigated. An example of such studies is provided by the work of Lee et al.[502].

The problems of spontaneous termination and of chain transfer deserve more thorough investigations. Unfortunately, not many data pertinent to that problem are available.

Anionic polymerization of lactons, lactams, Leuch's anhydrides, etc., deserves thorough discussion. References pertaining to some reviews covering this subject are given in the footnote on p. 6. The "activated monomer" mechanism is especially intriguing, although more conventional modes of those reactions should not be overlooked. The elementary steps of some lactone polymerization were recently investigated by Penczek and Slomkowski[503].

The choice of references is somewhat arbitrary and reflects the interest of this reviewer. Not surprisingly, the work done in his laboratory was emphasized.

Finally, I wish to thank the National Science Foundation for nearly 30 years' long support of our studies and to thank Professors Murray Goodman and Kurt Shuler for their hospitality in the Department of Chemistry of the University of California, San Diego, where this review was written.

VI. References

1. F. E. Matthews, E. H. Strange: Brit. Pat. *24,* 790 (1910)
2. G. Harries: Liebigs Ann. Chem. *463,* 1 (1928)
3. W. Schlenk, J. Appenrodt, A. Michael, A. Thal: Chem. Ber. *47,* 473 (1914)
4. H. Staudinger: ibid. *53,* 1073 (1920)
5. A. Wurz: C.r.hebd. Séanc. Acad. Sci. Paris; *86,* 1176 (1878)
6. K. Ziegler, H. Colonius, O. Schäfer: Liebigs Ann. Chem. *473,* 36 (1929); K. Ziegler, O. Schäfer: ibid. *479,* 150 (1930)
6a. K. Ziegler, L. Jabov, H. Wollthan, A. Wenz: ibid. *511,* 64 (1934)
7. K. Ziegler: Angew. Chem. *49,* 499 (1936)
8. W. Schlenk, E. Bergmann: Liebigs Ann. Chem. *464,* 1 (1928); *479,* 42, 58, 78 (1930)
9. W. Schlenk, J. Appenrodt, A. Michael, A. Thal: Chem. Ber. *47,* 473 (1914)
10. G. V. Schulz: Ergeb. Exakt. Naturw. *17,* 405 (1938)
11. J. L. Bolland: Proc. Roy. Soc. (London) A.*178,* 24 (1941)
12. A. Abkin, S. Medvedev: Trans. Faraday Soc. *32,* 286 (1936)
13. A. T. Blomquist, W. J. Tapp, J. R. Johnston: J. Amer. Chem. Soc. *67,* 1519 (1945)
14. R. G. Beaman: ibid. *70,* 3115 (1948)
15. R. E. Robertson, L. Marion: Can. J. Research, *26 B,* 657 (1948)
16. M. Szwarc, M. Levy, R. Milkovich: J. Amer. Chem. Soc. *78,* 2656 (1956); M. Szwarc: Nature *178,* 1168 (1956)
17. F. W. Stavely et al.: Ind. Engng. Chem. *48,* 778 (1956)
18. P. J. Flory: Principles of Polymer Chemistry, Cornell Univ. Press 1953
19. C. H. Bamford et al.: The Kinetics of Vinyl Polymerisation by Radical Mechanism, Butterworths, London 1958
20. H. Dostal, H. Mark: Z. Phys. Chem. B, *29,* 299 (1935)
21. P. J. Flory: J. Amer. Chem. Soc. *65,* 372 (1943)
22. M. Szwarc: Polymer Engeng. and Sci. *13,* 1 (1973)
23. K. Ziegler, K. Bähr: Chem. Ber. *61,* 253 (1928)
24. T. Aide, S. Inoue: Macromolec. *14,* 1162, 1166 (1981)
25a. N. Takada, S. Inoue: Makromol. Chem. *179,* 1377 (1978)
25b. S. Inoue, M. Koinuma, T. Tsuruta: ibid. *130,* 210 (1969)
26. Private communication from Prof. Inoue.
27. A. Rakatomanga, P. Hemery, S. Boileau, B. Lutz: Europ. Polymer. J. *14,* 581 (1978)
28. S. Penczek, P. Kubisa, K. Matyjaszewski; Adv. Polymer Sci. *37,* 1 (1980)
29. D. Vofsi, A. V. Tobolsky: J. Polymer. Sci. A*3,* 3261 (1965)
30. P. Dreyfuss, M. P. Dreyfuss: ibid. A*4,* 2179 (1966)
31. C. E. H. Bawn, R. M. Bell, A. Ledwith: Polymer *6,* 95 (1965)
32. G. Natta, I. Pasquon: Adv. Catal. *11,* 1 (1959)
33. G. Bier: Angew. Chem. *73,* 186 (1961)
34. E. G. Kontos, E. K. Eeasterbrook, R. D. Gilbert: J. Polymer Sci. *61,* 69 (1962)
35. Y. Doi, S. Neki, T. Keii: Macromolec. *12,* 814 (1979)
36. G. Natta, I. Pasquon, A. Zambelli: J. Amer. Chem. Soc. *84,* 1488 (1962)
37. M. E. Pruitt, J. M. Bagguett: U. S. Pat. 2,706,181 (1955)
38. R. C. Colclough, G. Gee: J. Polymer Sci. *34,* 171 (1959)

39. M. Osgan: ibid. A6, 1249 (1968)
40. R. Sakata, T. Tsuruta: Makromol. Chem. 40, 64 (1960)
41. E. J. Vandenberg: J. Polymer Sci. 47, 486 (1960)
42. M. Osgan, Ph. Teyssié: Polymer Lett. B5, 789 (1967)
43. Ph. Teyssié et al.: Inorg. Chim. Acta 19, 203 (1976)
44. E. J. Vandenberg: J. Polymer Sci. A7, 525 (1969)
45. A. Hamilton, R. Jerome, A. J. Hubert, Ph. Teyssié: Macromol. 6, 651 (1973)
46. Ph. Teyssié et al.: ACS Symp. Series 59, 165 (1977)
47. Huynh Ba Gia: Licence Thesis, Univ. Liege 1975
48. C. E. H. Bawn, A. Ledwith, P. Matthies: J. Polymer Sci. 24, 93 (1959)
49. A. J. Bloodworth, A. G. Davies: Proc. Chem. Soc. (London) A, p. 315, 1963
50. J. E. L. Roovers, S. Bywater: Trans. Faraday Soc. 62, 1876 (1966)
51a. J. E. Figueruelo, D. J. Worsfold: Europ. Polymer J. 4, 439 (1968)
51b. J. E. Figueruelo, A. Bello: J. Macromolec. Sci. 3, 311 (1969)
52. K. S. Kazanskii, A. A. Solovyanov, S. G. Entelis: Europ. Polymer J. 7, 1421 (1971)
53. C. C. Price, M. K. Akkapeddi: J. Amer. Chem. Soc. 94, 3972 (1972)
54. A. Deffieux, S. Boileau: Polymer 18, 1047 (1977)
55. A. Deffieux, E. Graff, S. Boileau: ibid. 22, 549 (1981)
56. B. De Groof, M. Van Beylen, M. Szwarc: Macromolec. 8, 396 (1975)
57. B. De Groof, W. Mortier, M. Van Beylen, M. Szwarc: ibid. 10, 598 (1977)
58. C. De Smedt, M. Van Beylen: ACS Symposiums Series, 166, 127 (1981)
59. C. Mathis, L. Christmann-Lamande, B. Francois: J. Polymer Sci. 16, 1285 (1978)
60. L. Christmann-Lamande, R. Nuffer, B. Francois: C.r. Acad. Sci. France, C280, 731, 941 (1975)
61. C. Mathis, L. Christmann-Lamande, B. Francois: Makromol. Chem. 176, 931 (1975)
62. C. Mathis, B. Francois: J. Polymer Sci. 16, 1297 (1978)
63a. S. Golden, C. Guttman, T. R. Tuttle: J. Amer. Chem. Soc. 87, 135 (1965)
63b. T. R. Tuttle, C. Guttman, S. Golden: J. Chem. Phys. 45, 2206 (1966)
64. J. W. Fletcher, W. A. Seddon: J. Phys. Chem. 79, 3055 (1975)
65. J. L. Dye, M. G. DeBacker, V. A. Nicely: J. Amer. Chem. Soc. 92, 5226 (1970)
66a. A. I. Popov: Pure and Appl. Chem. 51, 101 (1979)
66b. E. Mei, J. L. Dye, A. I. Popov: J. Amer. Chem. Soc. 99, 5308 (1977)
67. J. L. Dye et al.: ibid. 96, 608, 7203 (1974)
68. C. W. Kraus, W. W. Lucasse: ibid. 43, 2529 (1921)
69. J. P. Dodelet, F. Y. Jou, G. R. Freeman: J. Phys. Chem. 79, 2876 (1975)
70a. B. Bockrath, L. M. Dorfman: ibid. 77, 1002 (1973)
70b. G. A. Salmon, W. A. Seddon, J. W. Fletcher: Can. J. Chem. 52, 3259 (1974)
70c. G. A. Salmon, W. A. Seddon: Chem. Phys. Lett. 24, 366 (1974)
71a. S. Matalon, S. Golden, M. Ottolenghi: J. Phys. Chem. 73, 3098 (1969)
71b. M. T. Lok, F. J. Tehan, J. L. Dye: J. Phys. Chem. 76, 2975 (1972)
72. B. Bockrath, L. M. Dorfman: ibid. 79, 1509, 3064 (1975)
73a. W. L. Jolly: Adv. Chem. Ser. 50, 27 (1965)
73b. R. R. Dewald, R. B. Tsina: J. Phys. Chem. 72, 4520 (1968)
74. E. J. Kirschke, W. J. Jolly: Science 147, 45 (1965)
75. C. G. Overberger, E. M. Pearce, M. Mayers: J. Polymer Sci. 31, 217 (1958); 34, 109 (1959)
76. C. G. Overberger, H. Yuki, N. Urukawa: ibid. 45, 127 (1960)
77. F. S. Dainton, D. N. Wiles, A. N. Wright: J. Chem. Soc. p. 4283 (1960) and J. Polymer Sci. 45, 111 (1960)
78. C. G. Overberger, A. M. Schiller: J. Org. Chem. 46, 4230 (1961)
79. D. Laurin, G. Parravano: J. Polymer Sci. A6, 1047 (1968)
80. L. Christmann-Lamande, R. Nuffer, B. Francois: C.r. Acad. Sci. France, C280, 731, 941 (1975)
81. See for a review: M. Szwarc, J. Jagur-Grodzinski, in: Ions and Ion-Pairs in Organic Reactions, Vol. II, M. Szwarc (Ed.) John Wiley Publ. 1974
82. H. C. Wang, G. Levin, M. Szwarc: J. Amer. Chem. Soc. 100, 3969 (1978)
83. C. L. Lee, J. Smid, M. Szwarc: J. Phys. Chem. 66, 904 (1962)
84. A. Vrancken, J. Smid, M. Szwarc: Trans. Faraday Soc. 58, 2036 (1962)

85. D. H. Richards, R. L. Williams: J. Polymer Sci. *11*, 89 (1973)

86a. D. H. Richards: Polymer *19*, 109 (1978)

86b. D. H. Richards, in: Developments in Polymerisation, R. N. Howard (Ed.) Appl. Sci. Publ. (1979)

87. A. Davis, D. H. Richards, N. F. Scilly: Makromol. Chem. *152*, 121, 133 (1972)

88. A. V. Cunliffe, N. C. Paul, D. N. Richards, D. Thomson: Polymer *19*, 329 (1978)

89a. D. H. Richards, N. F. Scilly: Brit. Polymer J. *2*, 277 (1970)

89b. D. H. Richards et al.: ibid. *3*, 101 (1971)

90. D. H. Richards, N. F. Scilly: Chem. Comm. p. 1515 (1968)

91a. D. H. Richards et al.: Europ. Polymer J. *6*, 1469 (1970)

91b. A. V. Cunliffe, W. J. Hubbert, D. H. Richards: Makromol. Chem. *157*, 23, 39 (1972)

91c. D. H. Richards et al.: Polymer *16*, 654, 659, 665 (1975)

92. A. V. Tobolsky et al.: J. Polymer Sci. *28*, 425 (1958)

93a. A. V. Tobolsky, D. B. Hartley: ibid. A*1*, 15 (1963)

93b. D. B. George, A. V. Tobolsky: ibid. B*2*, (1964)

94. C. G. Overberger, N. Yamamoto: ibid. B*3*, 569 (1965); A*4*, 3101 (1964)

95. C. Mathis, B. Francois: Makromol. Chem. *156*, 7, 17 (1972)

96. H. C. Wang, G. Levin, M. Szwarc: J. Phys. Chem. *83*, 785 (1979)

97. R. Asami, M. Szwarc: J. Amer. Chem. Soc. *84*, 2269 (1962)

98. G. Spach, M. Monteiro, M. Levy, M. Szwarc: Trans. Faraday Soc. *58*, 1809 (1962)

99. C. E. Frank, W. E. Foster: J. Org. Chem. *26*, 303 (1961)

100. Z. Curos, P. Caluwe, M. Szwarc: J. Amer. Chem. Soc. *95*, 6171 (1973)

101. C. S. Marvel et al.: ibid. *61*, 2769, 2771 (1939)

102a. S. N. Khana, M. Levy, M. Szwarc: Trans. Faraday Soc. *58*, 747 (1962)

102b. R. Asami, S. N. Khana, M. Levy, M. Szwarc: ibid. *58*, 1821 (1962)

103a. J. Jagur, M. Levy, M. Feld, M. Szwarc: ibid. *58*, 2168 (1962)

103b. J. Jagur, M. Monteiro, M. Szwarc: ibid. *59*, 1353 (1963)

104. D. Gill, J. Jagur-Grodzinski, M. Szwarc: ibid. *60*, 1424 (1964)

105a. J. Jagur-Grodzinski, M. Szwarc: ibid. *59*, 2305 (1963)

105b. M. Szwarc: Proc. Roy. Soc. A*279*, 260 (1964); Ber. Bunsen-Ges. *67*, 763 (1963)

106. See for a review: M. Szwarc, A. Streitwieser, P. C. Mowery in: Ions and Ion-Pairs in Organic Reactions, M. Szwarc (Ed.) John Wiley Publ. 1974

107. D. H. Richards, M. Szwarc: Trans. Faraday Soc. *55*, 1644 (1959)

108. M. Morton, A. Rembaum, E. E. Bostick: J. Polymer Sci. *32*, 530 (1958)

109a. S. Boileau, G. Champetier, P. Sigwalt: ibid. C*16*, 3021 (1967)

109b. J. P. Favier, S. Boileau, P. Sigwalt: Europ. Polymer J. *4*, 3 (1968)

110. R. Asami, M. Levy, M. Szwarc: J. Chem. Soc. (London) p. 361 (1962)

111. W. C. E. Higginson, N. S. Wooding: ibid., p. 760 (1952)

112. P. Sigwalt, S. Boileau: J. Polymer Sci. C*62*, 51 (1978)

113. P. Sigwalt: ibid. C*50*, 95 (1975)

114. C. J. Chang, R. F. Kiesel, T. E. Hogen-Esch: J. Amer. Chem. Soc. *95*, 8446 (1973)

115. S. Searles, M. Tamres: Chem. Rev. *59*, 296 (1959)

116. S. Boileau, P. Hemery, B. Kaempf, F. Schué, M. Viguier: J. Polymer Sci., Polymer Lett. B*12*, 217 (1974)

117. S. Boileau, B. Kaempf, S. Raynal, J. Lacoste, F. Schué: ibid. B*12*, 214 (1974)

118. S. Boileau, B. Kaempf, J. M. Lehn, F. Schué: ibid. B*12*, 203 (1974)

119. See e.g. J. M. Lehn in: Structure and Bonding, Vol. 16, Springer-Verlag 1973

120. N. S. Wooding, W. C. E. Higginson: J. Chem. Soc. (London). p. 774 (1952)

121. e.g. B. Dietrich, J. M. Lehn: Tetrahedron Lett. *15*, 2025 (1973)

122. B. J. Schmitt, G. V. Schulz: Makromol. Chem. *121*, 184 (1969)

123. T. Shimomura, K. J. Tölle, J. Smid, M. Szwarc: J. Amer. Chem. Soc. *89*, 796 (1967)

124. K. Ziegler, K. Dislich: Chem. Ber. *90*, 1107 (1957)

125. R. E. Rundle: J. Phys. Chem. *61*, 45 (1957)

126a. D. Margerison, J. P. Newport: Trans. Faraday Soc. *59*, 2058 (1963)

126b. G. Wittig, F. J. Meyer, G. Lange: Liebigs Ann. Chem. *571*, 234 (1930)

127. T. L. Brown, M. T. Rogers: J. Amer. Chem. Soc. *79*, 1859 (1957)

128. M. Weiner, G. Vogel, R. West: Inorg. Chem. *1*, 654 (1962)

129. E. Warhurst: Disc. Faraday Soc. *2*, 239 (1947)
130. J. Berkowitz, D. A. Bafus, T. L. Brown: J. Phys. Chem. *65*, 1380 (1961)
131. T. L. Brown, J. A. Ladd: J. Organomet. Chem. *3*, 365 (1965)
132. E. Weiss, E. A. C. Lucken: ibid. *2*, 197 (1964)
133. T. L. Brown, D. W. Dickerhorf, D. A. Bafus: J. Amer. Chem. Soc. *84*, 1371 (1962)
134. T. L. Brown: Adv. Organomet. Chem. *3*, 365 (1965)
135. E. Weiss, G. Hencken: J. Organomet. Chem. *21*, 265 (1970)
136. H. Dietrich: Acta Cryst. *16*, 681 (1963)
137. T. L. Brown: Acc. Chem. Res. *1*, 23 (1968)
138. T. L. Brown, L. M. Seitz, B. Y. Kimura: J. Amer. Chem. Soc. *90*, 3245 (1968)
139. L. M. Seitz, T. L. Brown: ibid. *88*, 2174 (1966)
140. L. D. McKeever, R. Waack, M. A. Doran, E. B. Baker: ibid. *90*, 3224 (1968); *91*, 1057 (1969)
141. M. Szwarc: Carbanions, Living Polymers, and Electron Transfer Processes, Chapter IV, John Wiley Publ. 1968
142. I. Craubner: Z. Phys. Chem. *51*, 225 (1967)
143. P. West, R. Waack: J. Amer. Chem. Soc. *89*, 4395 (1967)
144. G. E. Hartwell, T. L. Brown: ibid. *88*, 4625 (1966)
145. H. C. Williams, T. L. Brown: ibid. *88*, 4134 (1966)
146. R. Waack, M. A. Doran, E. B. Baker; unpublished work, private communication by Dr. Waack
147. P. West, R. Waack: J. Amer. Chem. Soc. *89*, 4395 (1967)
148. M. Chabouel: J. Chim. Phys. *63*, 1143 (1966)
149. T. L. Brown, J. A. Ladd, G. N. Newman: J. Organomet. Chem. *3*, 1 (1965)
150. T. L. Brown et al.: J. Amer. Chem. Soc. *86*, 2135 (1964)
151. L. M. Seitz, T. L. Brown: ibid. *88*, 2174 (1966)
152. R. West, W. Glaze: ibid. *83*, 3580 (1961)
153. F. A. Settle, M. Haggerty, J. F. Eastham: ibid. *86*, 2076 (1964)
154 a. R. Waack, M. A. Doran, P. E. Stevenson: ibid. *88*, 2109 (1966)
154 b. G. Beinert: Bull. Soc. Chim. France; p. 3223 (1969)
155. C M. Selman, H. L. Hsieh: Polymer Lett. *9*, 219 (1971)
156. D. J. Worsfold, S. Bywater: Can. J. Chem. *38*, 1891 (1960)
157. K. F. O'Driscoll, A. V. Tobolsky: J. Polymer Sci. *35*, 259 (1959)
158. F. J. Welch: J. Amer. Chem. Soc. *81*, 1345 (1959)
159. S. Bywater, D. J. Worsfold: J. Organomet. Chem. *10*, 1 (1967)
160. W. H. Glaze, C. H. Freeman: J. Amer. Chem. Soc. *91*, 7198 (1969)
161 a. A. A. Solov'yanov, K. S. Kazanskii: Vysokomolek. Soedin A*12*, 2114 (1970)
161 b. K. S. Kazanskii, A. A. Solov'yanov, S. G. Entelis: Europ. Polymer J. *7*, 1421 (1971)
162. T. L. Brown: J. Organomet. Chem. *5*, 191 (1966)
163. K. Ziegler: Angew. Chem. *49*, 499 (1936)
164. H. L. Hsieh: J. Polymer Sci. A*3*, 163 (1965)
165. W. H. Glaze, C. M. Selman: J. Org. Chem. *33*, 1987 (1968)
166. E. N. Kropacheva, B. A. Dolgoplosk, E. M. Kuznetsova: Dokl. Akad. Nauk S.S.S.R. *130*, 253 (1960)
167 a. P. D. Bartlett, S. Friedman, M. Stiles: J. Amer. Chem. Soc. *75*, 1771 (1953)
167 b. P. D. Bartlett, S. J. Tauber, W. P. Weber: ibid. *91*, 6362 (1969)
168. A. G. Evans, D. B. George: J. Chem. Soc., p. 4653 (1961)
169. R. A. H. Casling, A. G. Evans, N. H. Rees: ibid. B, p. 519 (1966)
170. A. G. Evans, N. H. Rees: ibid. p. 6069 (1963)
171. M. Szwarc: J. Polymer Sci., Polymer Lett. *18*, 493 (1980)
172. J. E. L. Rovers, S. Bywater: Macromol. *8*, 251 (1975)
173. A. F. Johnson, D. J. Worsfold: J. Polymer Sci. A*3*, 449 (1965)
174. D. J. Worsfold, S. Bywater: Can. J. Chem. *42*, 2884 (1964)
175. J. E. L. Rovers, S. Bywater: Macromolec. *1*, 328 (1968)
176. F. Schué, S. Bywater: ibid. *2*, 458 (1969)
177. Unpublished results of Dr. S. Bywater
178. A. Guyot, J. Vialle: J. Macromolec. Sci. *4*, 79 (1970)
179. F. S. Dainton, K. J. Ivin: Nature *162*, 705 (1948); Quart. Rev. *12*, 61 (1958)

180. S. Bywater: Makromol. Chem. *52*, 120 (1960)
181. D. J. Worsfold, S. Bywater: J. Polymer Sci.
182. H. W. McCormick: ibid. *25*, 488 (1957)
183. A. Vrancken, J. Smid, M. Szwarc: Trans   2036 (1962)
184. K. J. Ivin, J. Leonard: Europ. Polymer
185. S. Bywater, D. J. Worsfold: J. Polym
186. M. Morton, R. F. Kammereck: J. A   17 (1970)
187. A. Roggero et al.: J. Polymer Sci.,   (1979)
188 a. M. P. Dreyfuss, P. Dreyfuss: Pol   ymer Sci. *4*, 2179 (1966)
188 b. C. E. H. Bawn, R. M. Bell, A.   1965)
189. K. J. Ivin, in: Reactivity, Mecha   ner Chemistry, A. D. Jenkins, A. Ledwith (Eds.) Wiley Publ. 197
190. J. Leonard, D. J. Maheux: J. M   1973)
191. S. Penczek, K. Matyjaszewski: ymp. *56*, 255 (1976)
192. C. L. Lee, J. Smid, M. Szwar   912 (1963)
193. K. J. Ivin, J. Leonard: Polyme
194. A. V. Tobolsky, W. J. MacKnight: Polymer Sulphur and Related Polymers; Intersc. 1965
195. W. B. Brown, M. Szwarc: Trans. Faraday Soc. *54*, 416 (1958)
196. A. Miyake, W. H. Stockmayer: Makromol. Chem. *88*, 90 (1965)
197. A. V. Tobolsky, A. Eisenberg: J. Colloid. Sci. *17*, 49 (1962)
198. E. J. Goethals: J. Polymer Sci., Polymer Symp. *56*, 271 (1976)
199. D. Van Ooteghem, E. J. Goethals: Makromol. Chem. *175*, 1513 (1974); *177*, 3389 (1976)
200. R. C. Schulz et al.: Amer. Chem. Soc. Symp. *59*, 77 (1977)
201. J. B. Rose: J. Chem. Soc., pp. 542, 546 (1956)
202. J. F. Brown, G. M. J. Slusarczuk: J. Amer. Chem. Soc. *87*, 931 (1965)
203. K. Jacobson, W. H. Stockmayer: J. Chem. Phys. *18*, 1600 (1950)
204. M. Szwarc, C. L. Perrin: Macromolec. *12*, 699 (1979)
205. P. Wittmer: Makromol. Chem. *103*, 188 (1967)
206 a. B. L. Funt, S. W. Laurent: Can. J. Chem. *42*, 2728 (1964)
206 b. B. L. Funt, D. Richardson, S. N. Bhadani: ibid. *44*, 711 (1966)
207. N. Yamazaki, S. Nakahama, I. Ianaka: Koggo Kagakn Zasshi, *70*, 1978 (1967)
208. S. N. Bhadani, G. Paravano: J. Polymer Sci. *8*, 225 (1970)
209. J. D. Anderson: ibid. A*6*, 3185 (1968)
210. S. Bywater, D. J. Worsfold: Can. J. Chem. *40*, 1564 (1962)
211. H. L. Hsieh, C. F. Wofford: J. Polymer Sci. A*7*, 449 (1969)
212. C. G. Buttler et al. ibid. A*16*, 937 (1978)
213. R. Kalir, A. Zilkha: Europ. Polymer J. *14*, 557 (1978)
214. V. E. J. Shashoua: J. Amer. Chem. Soc. *81*, 3156 (1959)
215. B. D. Coleman, T. G. Fox: ibid. *85*, 2141 (1963) and J. Polymer Sci. C*4*, 345 (1963)
216. R. V. Figini: Makromol. Chem. *71*, 193 (1964)
217. M. Szwarc, J. J. Hermans: J. Polymer Sci. B*2*, 815 (1964)
218. G. E. East, H. Furukawa: Polymer *20*, 659 (1979)
219. A. J. Bur, L. J. Fetters: Chem. Rev. *76*, 730 (1976)
220. P. J. Flory: J. Amer. Chem. Soc. *62*, 1561 (1940)
221 a. R. Waack, A. Rembaum, J. D. Coombs, M. Szwarc: ibid. *79*, 2026 (1957)
221 b. H. Brody, M. Ladacki, R. Milkovich, M. Szwarc: J. Polymer Sci. *25*, 221 (1957)
222. M. Szwarc: Adv. Chem. Phys. *2*, 147 (1959)
223. M. Litt, M. Szwarc: J. Phys. Chem. *62*, 508 (1958)
224. V. S. Nanda, R. K. Jain: J. Polymer Sci. A*2*, 4583 (1964)
225. M. Litt: J. Polymer Sci. *58*, 429 (1962)
226. W. T. Kynev, J. R. M. Radok, M. Wales: J. Chem. Phys. *30*, 363 (1959)
227. V. S. Nanda: Trans. Faraday Soc. *60*, 947, 949 (1964)
228. V. S. Nanda, S. C. Jain: Europ. Polymer J. *6*, 1605 (1970); *13*, 137 (1977)
229. T. A. Orofino, F. Wenger: J. Chem. Phys. *35*, 532 (1961)
230. R. V. Figini: Makromol. Chem. *44–46*, 497 (1961)
231. J. Largo-Cabrerizo, J. Guzman: Macromolec. *12*, 526 (1979)
232. B. A. Rozenberg et al.: Polymer Sci. USSR (English translation) *6*, 2253 (1964)

233a. R. V. Figini: Z. Phys. Chem. *23*, 224, 233 (1960)
233b. R. V. Figini, G. V. Schulz: Makromol. Chem. *41*, 1 (1960)
234. J. Grodzinsky, A. Katchalsky, D. Vofsi: ibid. *46*, 591 (1961)
235. D. Vofsi, A. Katchalsky: J. Polymer Sci. *26*, 127 (1957)
236. L. Horner, W. Jurgeleit, K. Klüpfel: Liebigs Ann. Chem. *591*, 108 (1955)
237. T. Ogawa, P. Quitana: J. Polymer Sci. A*13*, 2517 (1975)
238. V. Jaacks, C. D. Eisenbach, W. Kern: Makromol. Chem. *161*, 139 (1972)
239. T. Ogawa, T. Taninaka: J. Polymer Sci. A*10*, 2005 (1972)
240. V. Jaacks, N. Mathes: Makromol. Chem. *131*, 295 (1970)
241. V. Jaacks, G. Franzmann: ibid. *143*, 283 (1971)
242. N. Mathes, V. Jaacks: ibid. *142*, 209 (1971)
243. K. Boehlke, V. Jaacks: Angew. Chem. *81*, 336 (1969)
244. E. V. Kochetov, M. A. Markevitch, S. N. Enikolopyan: Dokledy Akad. Nauk SSSR *180*, 143 (1968)
245. Y. Etienne, R. Soulas: J. Polymer Sci. C*4*, 1061 (1964)
246. J. Trotman, M. Szwarc: Makromol. Chem. *37*, 39 (1960)
247a. T. Saegusa, S. Matsumoto: J. Polymer Sci. A*6*, 1559 (1968)
247b. T. Saegusa, S. Matsumoto, S. Kobayashi: Progress in Polymer Sci. Japan; *6*, 107 (1971)
248. Y. Yamashita, K. Ito, F. Nakakita: Makromol. Chem. *127*, 292 (1969)
249. T. Ogawa, J. Romero: Europ. Polymer J. *13*, 419 (1977)
250. D. S. Johnston, D. C. Pepper: Makromol. Chem. *182*, 393, 407, 421 (1981)
251. H. Gilbert et al.: J. Amer. Chem. Soc. *78*, 1669 (1956)
252. R. Gumbs, S. Penczek, J. Jagur-Grodzinski, M. Szwarc: Macromolec. *2*, 77 (1969)
253. J. Pac, P. H. Plesch: Polymer *8*, 252 (1967)
254. S. Penczek, J. Jagur-Grodzinski, M. Szwarc: J. Amer. Chem. Soc. *90*, 2174 (1968)
255. N. Oguni, M. Kamachi, J. K. Stille; Macromolec. *7*, 435 (1974)
256. M. P. Dreyfuss, J. C. Westfahl, P. Dreyfuss: ibid. *1*, 437 (1968)
257. R. A. Barzikina: Vysokomol. Soed. *16*, 906 (1974)
258. J. P. Schroeder, D. C. Schroeder, S. Yotikasthira: J. Polymer Sci. A*10*, 2189 (1972)
259. J. Poor, A. M. T. Finch: ibid. A*9*, 249 (1971)
260. J. W. Breitenbach, O. F. Olaj, F. Sommer: Adv. Polymer Sci. *9*, 48 (1972)
261. W. K. R. Barnikol, G. V. Schulz: Makromol. Chem. *68*, 211 (1963)
262. L. L. Böhm et al.: Adv. Polymer Sci. *9*, 1 (1972)
263. G. Geacintov, J. Smid, M. Szwarc: J. Amer. Chem. Soc. *83*, 1253 (1961); *84*, 2508 (1962)
264a. D. N. Bhattacharyya, C. L. Lee, J. Smid, M. Szwarc: Polymer *5*, 54 (1964)
264b. J. Phys. Chem. *69*, 612 (1965)
265a. G. V. Schulz et al.: Makromol. Chem. *71*, 198 (1964)
265b. Z. Phys. Chem. N.F. *45*, 269 (1965)
265c. ibid. *45*, 286 (1965)
266. V. Wartzelhan, G. Löhr, H. Höcker, G. V. Schulz: Makromol. Chem. *179*, 2211 (1978)
267a. W. K. R. Barnikol, G. V. Schulz: Z. Phys. Chem. N.F. *47*, 89 (1965)
267b. L. Böhm, W. K. R. Barnikol, G. V. Schulz: Makromol. Chem. *110*, 222 (1967)
268. F. S. Dainton, K. J. Ivin, R. T. LaFlair: Europ. Polymer J. *5*, 379 (1969)
269. R. V. Figini: J. Polymer Sci. C*16*, 2049 (1967); Makromol. Chem. *107*, 170 (1967)
270. D. J. Worsfold, S. Bywater: Can. J. Chem. *36*, 1141 (1958)
271a. G. Allen, G. Gee, C. Stretch: J. Polymer Sci. *48*, 189 (1960)
271b. C. Stretch, G. Allen: Polymer *2*, 151 (1961)
272. D. N. Bhattacharyya, J. Smid, M. Szwarc: J. Phys. Chem. *69*, 624 (1965)
273. F. S. Dainton, G. A. Harpell, K. J. Ivin: Europ. Polymer J. *5*, 395 (1969)
274. F. S. Dainton et al.: Makromol. Chem. *89*, 257 (1965)
275. J. Komiyama, L. L. Böhm, G. V. Schulz: ibid. *148*, 297 (1971)
276. Y. Gabillon, P. Piret: Acta Cryst. *15*, 1180 (1962)
277. T. A. Gribanova, T. Dniepropetr: Met. Inst. *44*, 217 (1961)
278. M. Shinohara: unpublished results of the Syracuse group
279. A. Parry, J. E. L. Roovers, S. Bywater: Macromolec. *3*, 355 (1970)
280. J. Smid, in: Ions and Ion-Pairs in Organic Reactions, Vol. I, p. 85; M. Szwarc (Ed.) Wiley Publ. 1972

281. T. Shimomura, K. J. Tölle, J. Smid, M. Szwarc: J. Amer. Chem. Soc. *89*, 796 (1967)
282. T. Shimomura, J. Smid, M. Szwarc: ibid. *89*, 5743 (1967)
283 a. C. Carvajal, K. J. Tölle, J. Smid, M. Szwarc: ibid. *87*, 5548 (1965)
283 b. T. E. Hogen-Esch, J. Smid: ibid. *88*, 318 (1966)
283 c. L. Böhm, G. V. Schulz: Ber. Bunsen-Ges. *73*, 260 (1969)
284. F. Accascina, R. M. Fuoss: Electrolytic Conductance, John Wiley Publ. 1965
285. D. J. Worsfold, S. Bywater: J. Chem. Soc., p. 5234 (1960)
286. D. N. Bhattacharyya, C. L. Lee, J. Smid, M. Szwarc: J. Phys. Chem. *69*, 608 (1965)
287. R. M. Fuoss: ibid. *68*, 903 (1964); *67*, 385, 1704 (1963)
288. T. E. Hogen-Esch, J. Smid: J. Amer. Chem. Soc. *88*, 307 (1966) and later publications
289. B. J. Schmitt, G. V. Schulz: (a) Makromol. Chem. *142*, 325 (1971); (b) Europ. Polymer J. *11*, 119 (1975)
290. G. Löhr, G. V. Schulz: Europ. Polymer J. *11*, 259 (1975)
291 a. T. Saegusa, H. Ikeda: Macromolec *5*, 354 (1972), *6*, 315 (1973)
291 b. T. Saegusa, S. Kobayashi, Y. Kimura: ibid. *7*, 1 (1974)
291 c. T. Saegusa et al.: ibid. *8*, 259 (1975)
291 d. T. Saegusa, J. Fuzukawa: ibid. *9*, 728 (1976), *10*, 73 (1977)
292. R. M. Fuoss: J.Amer. Chem. Soc. *57*, 488 (1935)
293. B. Funt, T. D. Williams: J. Polymer Sci. A *2*, 865 (1971)
294. G. Löhr, G. V. Schulz: Makromol. Chem. *77*, 264 (1964)
295. D. N. Bhattacharyya, J. Smid, M. Szwarc: J. Amer. Chem. Soc. *86*, 5024 (1964)
296 a. T. Saegusa, S. Kobayashi, Y. Kimuta: Pure and Appl. Chem. *48*, 307 (1976)
296 b. T. Saegusa: Ang. Chem. *89*, 867 (1977)
297. R. M. Fuoss, C. A. Krauss: J. Amer. Chem. Soc. *55*, 21, 476, 1019, 2387 (1933)
298. M. Eigen, K. Tamm: Z. Electrochem. *66*, 107 (1962)
299. M. Van Beylen, M. Fisher, J. Smid, M. Szwarc: Macromol. *2*, 575 (1969)
300. L. L. Böhm, G. V. Schulz: Makromol. Chem. *153*, 5 (1972)
301. G. Löhr, S. Bywater: Can. J. Chem. *48*, 2031 (1970)
302. B. J. Schmitt, G. V. Schulz: Makromol. Chem. *175*, 3261 (1974)
303. M. Chmelir, G. V. Schulz: Ber. Bunsen-Ges. *75*, 830 (1971)
304. L. L. Böhm, G. V. Schulz: Ber. Bunsen-Ges. *73*, 260 (1969)
305. L. L. Böhm, G. Löhr, G. V. Schulz: Ber. Bunsen-Ges. *78*, 1064 (1974)
306. J. M. Alvarino, M. Chmelir, B. J. Schmitt, G. V. Schulz: J. Polymer Sci. C *42*, 155 (1973)
307. J. Comyn, F. S. Dainton, K. J. Ivin: Europ. Polymer J. *6*, 319 (1970)
308. M. Bunge, G. Löhr, H. Höcker, G. V. Schulz: Europ. Polymer J. *13*, 283 (1977)
309. L. V. Vinogradova, V. N. Zgonnik, N. I. Nikolaev. K. H. Tsvetsanov: Europ. Polymer J. *15*, 545 (1979)
310. G. Löhr, G. V. Schulz: Makromol. Chem. *117*, 283 (1968)
311. R. Fletcher, M. J. D. Powell: Comput. J. *6*, 163 (1963)
312. F. S. Dainton, K. M. Hui, K. J. Ivin: Europ. Polymer J. *5*, 387 (1969)
313. J. Comyn: Ph. D. Thesis, Univ. Leeds (1967)
314. M. Shima, D. N. Bhattacharyya, J. Smid, M. Szwarc: J. Amer. Chem. Soc. *85*, 1306 (1963)
315. R. V. Figini: Makromol. Chem. *107*, 170 (1967)
316. D. N. Bhattacharyya, C. L. Lee, J. Smid, M. Szwarc: J. Amer. Chem. Soc. *85*, 533 (1963)
317. H. Sadek, R. M. Fuoss: ibid. *76*, 5897, 5905 (1954)
318. S. Winstein, E. Clippinger, A. K. Feinberg, G. P. Robinson: ibid. *76*, 2597 (1954)
319. R. V. Slater, M. Szwarc: ibid. *89*, 6043 (1967)
320. L. L. Chan, J. Smid: ibid. *90*, 4654 (1968)
321. K. L. Tölle, J. Smid, M. Szwarc: J. Polymer Sci. B *3*, 1037 (1965)
322. L. L. Böhm, M. Chmelir, G. Löhr, B. J. Schmitt, G. V. Schulz: Adv. Polymer Sci. *9*, 1 (1972)
323 a. M. Szwarc: Makromol. Chem. *89*, 44 (1965)
323 b. M. Szwarc: Ions and Ion-Pairs, in: Carbanions, Living Polymers and Electron Transfer Processes; Intersc. Publ. 1968
323 c. Forward by M. Szwarc, in: Ions and Ion-Pairs in Organic Chemistry, Vol. II, M. Szwarc (Ed.) John Wiley Publ. 1974
324. P. Chang, R. V. Slater, M. Szwarc: J. Phys. Chem. *70*, 3180 (1966)
325. T. E. Gough, P. R. Hindle: Can. J. Chem. *47*, 1698, 3393 (1969)

326. M. Shinohara, J. Smid, M. Szwarc: J. Amer. Chem. Soc. *90*, 2175 (1968)

327. M. Van Beylen, D. N. Bhattacharyya, J. Smid, M. Szwarc: J. Phys. Chem. *70*, 157 (1966)

328. D. J. Worsfold, S. Bywater: ibid. *70*, 162 (1966)

329. J. Kirkwood: J. Chem. Phys. *24*, 233 (1939)

330a. E. Grunwald, A. L. Bacarella: J. Amer. Chem. Soc. *80*, 3840 (1958)

330b. E. Grunwald, G. Baughman, G. Kohnstam: ibid. *82*, 5801 (1960)

331. M. Shinohara, J. Smid, M. Szwarc: Chem. Communications, 1232 (1969)

332. K. J. Laidler, H. Eyring: Ann. N. Y. Acad. Sci. *39*, 303 (1940)

333. K. Strehlow, W. Knoche, H. Schneider: Ber. Bunsen-Ges. *77*, 761 (1973)

334. P. Debye: Trans. Electrochem. Soc. *82*, 265 (1942)

335a. A. Persoons: J. Phys. Chem. *78*, 1210 (1974)

335b. L. DeMaeyer, A. Persoons: Techniques of Chemistry Vol. II, Part 2, p. 211; Weissberger, Hemmes (Eds.) Intersc. Publ. 1973

336. A. Persoon, J. Everaet, in: Anionic Polymerization, ACS Symposium *166*, 153 (1981) J. E. McGrath (Ed.)

337. L. L. Böhm: Z. Phys. Chem. N. F. *72*, 199 (1970)

338. N. M. Atherton, S. I. Weissman: J. Amer. Chem. Soc. *83*, 1330 (1961)

339. N. Hirota: ibid. *90*, 3603 (1968)

340. C. L. Dodson, A. H. Redoch: J. Chem. Phys. *48*, 3226 (1968)

341a. N. Hirota, R. Kreilick: J. Amer. Chem. Soc. *88*, 614 (1966)

341b. A. Crowley, N. Hirota, R. Kreilick: J. Chem. Phys. *46*, 4815 (1967)

342. H. van Willengen, J. A. M. van Broekhoven, E. de Boer: Molec. Phys. *12*, 533 (1967)

343. J. H. Sharp, M. C. R. Symons, in: Ions and Ion-Pairs in Organic Reactions Vol I., M. Szwarc (Ed.) Wiley-Intersc. Publ. 1972

344. K. Höfelmann, J. Jagur-Grodzinski, M. Szwarc: J. Amer. Chem. Soc. *91*, 4645 (1969)

345. W. J. Le Noble, A. R. Das: J. Phys. Chem. *74*, 3429 (1970)

346a. S. Claesson, B. Lundgren, M. Szwarc: Trans. Faraday Soc. *66*, 3053 (1970)

346b. B. Lundgren, S. Claesson, M. Szwarc: Chemica Scripta *3*, 49 (1973)

346c. B. Lundgren, S. Claesson, M. Szwarc: ibid. *3*, 53 (1973)

347. M. Bunge, H. Höcker, G. V. Schulz: Makromol. Chem. *180*, 2637 (1979)

348. T. Aide, S. Inoue: Macromol. *14*, 1162, 1166 (1981)

349. B. C. Anderson et al.: ibid. *14*, 1599 (1981)

350. L. Vancea, S. Bywater: ibid. *14*, 1321 (1981)

351. R. F. Adams, T. L. Staples, M. Szwarc: Chem. Phys. Lett. *5*, 474 (1970)

352. D. Honnore, J. C. Favier, P. Sigwalt, M. Fontanille: Europ. Polymer J. *10*, 425 (1974)

353. M. Fisher, M. Szwarc: Macromol. *3*, 23 (1970)

354. M. Tardi, D. Rouge, P. Sigwalt: Europ. Polymer J. *3*, 85 (1967)

355. M. Tardi, P. Sigwalt: Europ. Polymer J. *8*, 151 (1972)

356. A. Soum, M. Fontanille, P. Sigwalt: J. Polymer Sci. A*15*, 659 (1977)

357. R. V. Slater, M. Szwarc: J. Amer. Chem. Soc. *89*, 6043 (1967)

358. L. L. Chan, J. Smid: ibid. *89*, 4547 (1967)

359. G. Löhr, G. V. Schulz: Makromol. Chem. *77*, 240 (1964)

360a. A. Roig, J. E. Figueruelo, E. Llano: J. Polymer Sci. B*3*, 171 (1965); C*16*, 4141 (1968)

360b. G. M. Guzman, A. Bello: Makromol. Chem. *107*, 46 (1967)

361. A. Rembaum, M. Szwarc: J. Polymer Sci. *22*, 189 (1956)

362. T. G. Fox et al.: J. Amer. Chem. Soc. *80*, 1768 (1958)

363. D. L. Glusker, E. Stiles, B. Yoncoskie: J. Polymer Sci. *49*, 297 (1961)

264. S. Bywater: Adv. Polymer Sci. *4*, 66 (1965)

365. G. Löhr, G. V. Schulz: Makromol. Chem. *172*, 137 (1973)

366. G. Löhr, B. J. Schmitt, G. V. Schulz: Z. Phys. Chem. *78*, 177 (1972)

367. G. Löhr, G. V. Schulz: ibid. *65*, 170 (1969)

368. J. E. Figueruelo: Makromol. Chem. *131*, 63 (1970)

369. G. Löhr, G. V. Schulz: Europ. Polymer J. *10*, 121 (1974)

370. I. Mita, Y. Watanabe, T. Akatsu, H. Kambe: Polymer J. *4*, 271 (1973)

371. J. Terkoval, P. Kratochril: J. Polymer Sci. A*10*, 1391 (1972)

372. L. Lochmann, M. Rodava, J. Petranek, P. Lim: ibid. A*12*, 2295 (1974)

373. V. Warzelhan, H. Höcker, G. V. Schulz: Makromol. Chem. *179*, 2221 (1978)

374. V. Warzelhan, H. Höcker, G. V. Schulz: ibid. *181*, 149 (1980)
375. V. Warzelhan, G. V. Schulz: ibid. *177*, 2185 (1976)
376. M. Szwarc: see Ref. [141], p. 498
377. F. M. Brower, H. W. McCormick: J. Polymer Sci. A*1*, 1749 (1963)
378. R. P. Chaplin, S. Yaddehigi, W. Ching: Europ. Polymer J. *15*, 5 (1979)
379. R. Kraft, A. H. E. Müller, V. Warzelhan, H. Höcker, G. V. Schulz: Macromol. Chem. *11*, 1093 (1978)
380. W. Fowells, C. Schuerch, F. A. Bovey, F. P. Hood: J. Amer. Chem. Soc. *89*, 1396 (1967)
381. F. A. Bovey: High Resolution NMR of Macromolecules, Academic Press 1972
382. R. K. Graham, D. L. Dunkelberger, E. S. Cohn: J. Polymer Sci. *42*, 501 (1960)
383. R. Kraft, A. H. E. Müller, H. Höcker, G. V. Schulz: Makromol. Chem., Rapid Comm. *1*, 363 (1980)
384. A. H. E. Müller: ACS Symp. Series *160*, 441 (1981)
385. S. N. Khanna, M. Levy, M. Szwarc: Trans. Farady Soc. *58*, 747 (1962)
386. L. Vancea, S. Bywater: preprint of 26th IUPAC Meeting, Mainz, p. 140, 1979
387. P. E. M. Allen, M. C. Fisher, C. Mair, E. H. Williams: ACS Symp. Series 166, 185 (1981)
388. M. Morton, E. E. Bostick, R. Livigny: Rubber and Plastic Age *42*, 397 (1961)
389. D. J. Worsfold, A. F. Johnston, S. Bywater: Can. J. Chem. *42*, 1255 (1964)
390. Yu. L. Spivin, A. R. Gantmakher, S. S. Medvedev: Dokl. Akad. Nauk SSSR *146*, 368 (1962)
391. H. Sinn, F. Patat: Angew. Chem. *75*, 805 (1963)
392. D. J. Worsfold, S. Bywater: Macromolec. *5*, 393 (1972)
393. S. Bywater, D. J. Worsfold: Can. J. Chem. *45*, 1821 (1967)
394 a. A. Garton, S. Bywater: Macromolec. *8*, 697 (1975)
394 b. A. Garton, S. Bywater: ibid. *8*, 694 (1975)
395. B. Funt, V. Hornoff: J. Polymer Sci. A-2 *9*, 2429 (1971)
396. H. C. Wang, M. Szwarc: Macromolec *15*, 208 (1982)
397. M. Szwarc: ACS Symp. Series *166*, 1 (1981)
398. A. Gourdenne, P. Sigwalt: Europ. Polymer. J. *3*, 481 (1967)
399. A. Siove, P. Sigwalt, M. Fontanille: Polymer *16*, 605 (1975)
400. A. Garton, R. P. Chaplin, S. Bywater: Europ. Polymer J. *12*, 697 (1976)
401. C. J. Dyball, D. J. Worsfold, S. Bywater: Macromolec. *12*, 819 (1979)
402. Z. Laita, M. Szwarc: ibid. *2*, 412 (1969)
403. A. Yamagishi, M. Szwarc: ibid. *11*, 504 (1978)
404. F. Schué, D. J. Worsfold, S. Bywater: ibid. *3*, 509 (1970)
405. S. Brownstein, S. Bywater, D. J. Worsfold: ibid. *6*, 715 (1973)
406. F. Bandermann, H. Sinn: Makromol. Chem. *96*, 150 (1966)
407. G. Beinert, P. Lutz, E. Franta, P. Rempp: ibid. *179*, 551 (1978)
408. K. F. O'Driscoll, R. Patsiga: J. Polymer Sci. A*3*, 1037 (1965)
409. A. Yamagishi, M. Szwarc: Macromolec. *11*, 1091 (1978)
410. A. A. Korotkov: Angew. Chem. *70*, 85 (1958)
411. K. F. O'Driscoll, I. Kuntz: J. Polymer Sci. *61*, 19 (1962)
412. M. Morton, F. R. Ells: J. Polymer Sci. *61*, 25 (1962)
413. A. F. Johnson, D. J. Worsfold: Makromol. Chem. *85*, 273 (1965)
414. A. Yamagishi, M. Szwarc, L. Tung, G. Y-S. Lo: Macromolec. *11*, 607 (1978)
415. D. J. Worsfold: J. Polymer Sci. *20*, 2783 (1967)
416. L. H. Tung, G. Y-S. Lo, D. E. Boyer: Macromolec. *11*, 616 (1978)
417. P. Lutz, G. Beinert, E. Franta, P. Rempp: Europ. Polymer. J. *15*, 1111 (1979)
418. S. Bywater, D. J. Worsfold: Can. J. Chem. *40*, 1564 (1962)
419. D. J. Worsfold: J. Polymer Sci. *20*, (Polymer Physics); 99 (1982)
420. M. Delamar et al.: ibid. A*20*, 245 (1982)
421. E. Loria et al.: Polymer *22*, 95, 1419 (1981)
422. J. E. Figueruello, M. Rodriguez: Makromol. Chem. *176*, 3107 (1975)
423. J. E. Figueruello, S. Nenna: Europ. Polymer J. *11*, 511 (1975)
424. L. P. Blanchard: J. Polymer Sci. A*10*, 1353, 3082 (1972)
425. S. Boileau, G. Champetier, P. Sigwalt: Makromol. Chem. *69*, 180 (1963)
426. G. Tersae, S. Boileau, P. Sigwalt: ibid. *149*, 153 (1972)
427. P. Hemery, S. Boileau, P. Sigwalt: J. Polymer Sci. Symp. *52*, 189 (1975)

428. C. C. Price, M. K. Akkapeddi, B. F. DeBona, B. C. Furie: J. Amer. Chem. Soc. *94*, 3964 (1972)
429. K. Matsuzaki, H. Ito: J. Polymer Sci. A*15*, 647 (1977)
430. P. Sigwalt: Pure and Applied Chem. *48*, 257 (1976)
431. J. Geerts, M. Van Beylen, G. Suets: J. Polymer Sci. A*7*, 2859 (1969)
432. J. E. L. Roovers, S. Bywater: Trans. Faraday Soc. *62*, 701 (1962)
433. J. E. L. Roovers, S. Bywater: Polymer *14*, 594 (1973)
434. G. E. M. Jones, O. L. Highes: J. Chem. Soc. (London) p. 1197 (1934)
435. C. E. H. Bawn, A. Ledwith, N. McFarlane: Polymer *10*, 653 (1969)
436. J. E. Figueruello, S. Nenna: Makromol. Chem. *176*, 3377 (1975)
437. J. H. Exner, E. C. Steiner: J. Amer. Chem. Soc. *96*, 1782 (1974)
438. C. C. Price, D. D. Carmelite: ibid. *88*, 4039 (1966)
439. G. Beinert, P. Lutz, E. Franta, P. Rempp: Makromol. Chem. *179*, 551 (1978)
440. P. Guyot, J. C. Favier, M. Fontanille, P. Sigwalt: Polymer *22*, 1724 (1981); *23*, 73 (1982)
441. H. Tani, N. Oguni: Polymer Lett. *7*, 803 (1969)
442. E. Graf, J. M. Lehn: J. Amer. Chem. Soc. *97*, 5022 (1975)
443. G. Ndebeka, P. Canber, S. Reynal, S. Lecoliere: Polymer *22*, 347, 356 (1981)
444. H. Brody, D. H. Richards, M. Szwarc: Chem. and Ind. p. 1473 (1958)
445. A. A. Soloyyanov, K. S. Kazanskii: Vyskomol. Soedin (A) *14*, 1063, 1071 (1972)
446. G. Merle et al.: J. Polymer Sci. A*15*, 2067 (1977)
447. J. P. Pascault, J. Kawak, J. Gole, Q. T. Pham: Europ. Polymer J. *10*, 1107 (1974)
448. C. C. Price: Acc. Chem. Res. *7*, 294 (1974)
449. M. Szwarc: Pure Appl. Chem. *48*, 247 (1976)
450. P. Hemery, S. Boileau, P. Sigwalt: Europ. Polymer J. *7*, 1581 (1971)
451. G. Mengoli, S. Daolio: Polymer Lett. *13*, 743 (1975)
452. C. A. Uraneck, J. N. Short, R. P. Zelinsky: U.S. Pat. *3*, 135, 716 (1966)
453. Y. Nitadori, E. Franta, P. Rempp: Makromol. Chem. *179*, 927 (1978)
454. P. Chaumont, J. Herz, P. Rempp: Europ. Polymer J. *15*, 537 (1979)
455. J. Brossas, G. Clouet: C. R. Acad. Sci. C-*280*, 1459 (1975)
456. R. Milkovich: ACS Symp. Series *166*, 47 (1981)
457. D. H. Richards: Brit. Polymer J. *12*, 89 (1980)
458. F. J. Burgess et al.: Polymer *18*, 719, 726, 733 (1977)
459. J. Lehman: to be published
460. F. J. Burgess et al.: Polymer *19*, 334 (1978)
461. P. Cohen, F. Schué, D. H. Richards: to be published
462. G. Spach, M. Levy, M. Szwarc: J. Chem. Soc. (London) *1962*, 355
462a. M. Szwarc: Makromol. Chem. *35* A, 123 (1960)
463. S. Boileau and P. Sigwalt: unpublished work
464a. M. Katayama et al.: Bull. Chem. Soc. Jpn, *38*, 851 and 2208 (1965)
464b. D. J. Metz, R. C. Potter, J. K. Thomas; J. Polymer Sci, A1, *5*, 877 (1967)
464c. C. Schneider, A. J. Swallow: Makromol. Chem. *114*, 155 (1968)
465. T. E. Hogen-Esch, M. J. Plodinec: J. Phys. Chem. *80*, 1090 (1976)
466. M. Yamaoka, F. Williams, K. Hayashi: Trans. Faraday Soc. *63*, 376 (1967)
467. P. Kebarle, in: Ions and Ion-Pairs in Organic Reactions, Vol. I, pp. 67–74, M. Szwarc (Ed.) John Wiley Publ. 1972
468. H. Hsieh: a paper read in March 1982 meeting in ACS (Las Vegas) and presently in the process of publication
469a. P. J. Flory: J. Amer. Chem. Soc. *61*, 1518 (1939)
469b. F. T. Wall: ibid. *62*, 803 (1940); *63*, 821 (1941)
470a. G. Natta: J. Polymer Sci. *16*, 143 (1955)
470b. G. Natta et al.: J. Amer. Chem. Soc. *77*, 1708 (1955)
471. F. A. Bovey, G. V. D. Tiers: J. Polymer Sci. *44*, 173 (1960)
472. T. Yoshino, J. Komiyama: ibid. B, *3*, 311 (1965)
473. M. Morton, R. A. Pett, L. J. Fetters: Macromolec. *3*, 334 (1970)
474. M. Szwarc: Macromolec. *15*, 1449 (1982)
475. L. H. Tung et al.: U.S. Patent 4, 196, 153 (1980).
476. P. Cohen, M. J. M Abadie, F. Schué, D. H. Richards: Polymer *23*, 1348, 1105 (1982)

477. F. A. Bovey: Polymer Conformation and Configuration, Academic Press 1969
478. S. Bywater, D. J. Worsfold, G. Hollingsworth: Macromolec. *5*, 389 (1972)
479 a. D. J. Cram, K. R. Kopecky: J. Amer. Chem. Soc. *81*, 2748 (1959)
479 b. D. J. Cram: J. Chem. Education *37*, 317 (1960)
480 a. B. S. Garret et al.: J. Amer. Chem. Soc. *81*, 1007 (1959)
480 b. T. J. Loitereg, D. J. Cram: ibid. *90*, 4011, 4019 (1968)
481. I. C. Randall, F. E. Naylor, H. L. Hsieh: Macromolec. *3*, 497 (1970)
482. D. Braun et al.: Makromol. Chem. *51*, 15 (1962)
483. H. S. Makowski, M. Lynn: J. Macromol. Chem. *1*, 443 (1966)
484 a. B. D. Coleman, T. G. Fox: J. Chem. Phys. *38*, 1065 (1963)
484 b. B. D. Coleman, T. G. Fox: J. Polymer Sci. A*1*, 3183 (1963)
485. W. H. Glaze et al.: J. Organometal. Chem. *44*, 39 (1971)
486. S. Bywater, D. J. Worsfold: ibid. *159*, 229 (1978)
487. S. Bywater: ACS Symposium Series *166*, 71 (1981)
488. S. J. Kennedy, J. C. Wheeler: J. Chem. Phys. (in press)
489 a. J. C. Wheeler, S. J. Kennedy, P. Pfeuty: Phys. Rev. Lett. *45*, 1768 (1980)
489 b. J. C. Wheeler, P. Pfeuty: Phys. Rev. A*24*, 1050 (1981)
490. J. C. Wheeler, P. Pfeuty: Phys. Rev. Lett. *46*, 1409 (1981); J. Chem. Phys. *74*, 6415 (1981)
491. L. B. Ebbert: J. Molec. Catalysts
492. M. Goodman, E. Peggion: Pure and Appl. Chem. *53*, 699 (1981)
493. Y. Shalitin: in "Ring Opening Polymerisation", K. P. Frisch and S. L. Reegen, Eds, Marcel Dekker, Publ. (1969)
494. M. K. Reimschuessel: in "Ring...
495. R. D. Lundberg, E. F. Cox: in "Ring...
496. M. Szwarc: Advan. Polymer Sci. *4*, 1 (1965)
497. T. Saegusa et al.: Macromolec. *6*, 657 (1973)
498 a. M. Katayama et al.: Bull. Chem. Soc. Jpn. *38*, 851, 2208 (1965)
498 b. D. J. Metz, R. C. Potter, J. K. Thomas: J. Polymer Sci. *5*, 877 (1967)
498 c. C. Schneider: Adv. Chem. Ser. *91*, 219 (1968)
499. M. Van Beylan and B. De Groof: Makromolek. Chem. *179*, 1814 (1978)
500. M. Shima et al.: J. Amer. Chem. Soc. *85*, 533 (1963)
501 a. F. Bhasteter, J. Smid, M. Szwarc: ibid. *85*, 3909 (1963)
501 b. J. Stearne, J. Smid, M. Szwarc: Trans. Faraday Soc. *60*, 2054 (1964)
502. C. L. Lee, J. Smid, M. Szwarc: ibid. *59*, 1192 (1963)
503. T. Saegusa: in "Macromolecules" p. 31, Ed. H. Benoit and P. Rempp, Pergamon Press (1982)
504. S. Penczek and S. Slomkowski: ACS Symposium Ser. *166*, 271 (1981)

Received September 1st, 1982

Subject Index

Author Index Volumes 1–49

Advances in Polymer Science

Fortschritte der
Hochpolymeren-Forschung

Editors:
H.-J. Cantow, G. Dall'Asta,
K. Dušek, J. D. Ferry,
H. Fujita, M. Gordon,
J. P. Kennedy, W. Kern,
S. Okamura,
C. G. Overberger,
T. Seagusa, G. V. Schulz,
W. P. Slichter, J. K. Stille

Springer-Verlag
Berlin
Heidelberg
New York

Volume 34/35
A. Gandini, H. Cheradame

Cationic Polymerisation

Initiation Processes with Alkenyl Monomers

1980. 12 figures, 9 tables. X, 289 pages
ISBN 3-540-10049-0

This monograph covers the entire spectrum of initiation systems in the cationic polymerisation of alkenyl monomers. Following a detailed outline of the factors which play an important role in determining the behaviour of cationic polymerisation, each type of initiation is discussed individually. Particular emphasis is placed on the two major modes of initiation: initiation by Brønsted acids and initiation by Lewis acids. The authors analyze the present status of this discipline through a critical review of the literature and a series of specific mechanistic proposals, some of which are entirely new. Published material relevant to the understanding of the processes leading to the formation and characterisation of active species is covered exhaustively. The significance of early work is reinterpreted and the impact of more recent studies as well as their shortcomings assessed. The potentials of new experimental techniques are also discussed. Finally, suggestions are offered for future work in many areas on the basis of the mechanistic proposals developed.
This book will help stimulate further ideas, discussions and research in a discipline which is experiencing a lively renaissance. (904 references)

Contents: Introduction. – Fundamentals. – Initiation by Brønsted Acids and Iodine. – Initiation by Lewis Acids. – Initiation by Carbenium Salts and Related Species. – Initiation by Bare Cations. – Electrochemical Initiation. – Photoinitiation. – Initiation from a Charge-Transfer Complex. – Initiation from a Polymer. – Miscellaneous Initiators. – References. – Subject Index.

Advances in Polymer Science

Fortschritte der Hochpolymeren-Forschung

Editors:
H.-J. Cantow, G. Dall'Asta,
K. Dušek, J. D. Ferry,
H. Fujita, M. Gordon,
J. P. Kennedy, W. Kern,
S. Okamura,
C. G. Overberger,
T. Seagusa, G. V. Schulz,
W. P. Slichter, J. K. Stille

Springer-Verlag
Berlin
Heidelberg
New York

Volume 37
S. Penczek, P. Kubisa, K. Matyjaszewski

Cationic Ring-Opening Polymerization of Heterocyclic Monomers

1980. 19 figures. V, 156 pages
ISBN 3-540-10209-4

The detailed understanding of the chemistry of the elementary reactions of cationic polymerization, including the corresponding kinetic parameters, is only available for the ring-opening polymerization of heterocyclics. In this volume, the authors present a modern review of cationic polymerization based on the results obtained primarily during the last decade. Although discussion is centered mainly on the polymerization of heterocyclic monomers, some general rules are formulated. Elementary reactions are treated separately; quantitative data were carefully selected to present the relationship between structure and reactivity. For deeper insights into the problems related to cationic polymerization, this review is organized such that similar *mechanisms* rather than similar groups of monomers are trested. The ways in which cationic polymerization of heterocyclic monomers can be initiated, the mechanisms of growth of polymer chain, involving related equilibria and reactions responsible for termination of material and kinetic chain, are analysed from the point of view of their similarities and differences. The review not only summarises the present state of knowledge but also indicates future trends in studies of cationic polymerization. (282 references)

Contents: Introduction. – Monomer Structures, Ring Strains and Nucleophilicities (Basicities). – Initiation. – Propagation. – Termination and Transfer Processes. – Addendum. – References. – Subject Index.